ET CONTACT

BLUEPRINT FOR A NEW WORLD

BY
GINA LAKE

Oughten House Publications
P.O. Box 2008 Livermore,
CA 94551-2008 USA
Phone (510) 447-2332 Fax (510) 447-2376
Email: oughtenhouse@rest.com
www.oughtenhouse.com

ET Contact: Blueprint for a New World

by

Gina Lake

Copyright © 1997 Gina Lake

Published 1997.

00 99 98 97   0 9 8 7 6 5 4 3 2 1

All rights reserved. This book may not be reproduced, stored in a retrieval system, or transmitted in any form by electronic, video, laser, mechanical, photocopying, recording means or otherwise, in part or in whole, without written permission from the author.

Cover art by David Adams

Published by

OUGHTEN HOUSE PUBLICATIONS

PO BOX 2008

LIVERMORE, CALIFORNIA 94551 USA

Phone: (510) 447-2332

FAX: (510) 447-2376

E-MAIL: oughtenhouse.com

www.oughtenhouse.com

Library of congress cataloging in publication

Lake, Gina, 1951-
    ET contact : blueprint for a new world / by Gina Lake.
       p.   cm.
    ISBN 1-880666-62-6
       1. Spirit writings.   2. Life on other planets--Miscellanea.
I. Title.
BF1311.L43L35   1997
133.9'3--dc21

                                        96-52117
                                            CIP

ISBN 1-880666-62-6

PRINTED IN THE USA

Contents

Acknowledgments v

Dedication vii

Preface viii

Introduction ix

1. Who Is Here 1
   What's an Extraterrestrial?—Two Varieties of Extraterrestrial—Star People and Walk-Ins—Fourth-density Beings—Fifth-density Beings—Sixth-density Beings—Who's Been Here

2. What We Are Doing Here 13
   Healing and Teaching—Guiding—The Abductions—Hybridization—Peace—Ecology—Negative Extraterrestrials

3. How Star People Are Helping Us 25
   Star People in Education—Star People in Politics—Star People in the Creative Arts—Star People in the Media—Star People in Science and Medicine—Star People as Healers, Psychics, and Channels

4. Your Changing World 39
   Famine, Pestilence, and Disease—Climate Changes—Geological Changes—Pole Shift—Coping With Change

5. The Role of Evil in Your World 47

6. Our Plan and How We Operate 53

7. Life With ET's 63
   Building a New World Together—Life With ET's—The Challenges of Integrating ET's into Your World—The Impact of ET's on Your World—The World of the Future

8. What You Can Do to Help 75
   What You Can Do to Help Yourself—What You Can Do to Help the Planet—What You Can Do to Help Us

9. On the Way to Ascension 83
   What Is Ascension?—Preparing for Ascension

ALSO BY THE AUTHOR

Pathways to Self Discovery: Tools to Help You Access Your Higher Self for Guidance and Healing

The Extraterrestrial Vision: The ET Agenda, with Theodore

FORTHCOMING BOOKS FROM THE SAME AUTHOR

Wake Up!

Eight Steps to God

ACKNOWLEDGMENTS

Many thanks to visionary artist, David Adams, and his wife Trisha for the beautiful cover image he created for this book and so generously contributed. His work captures the wonderful awe of contact with new worlds.

ABOUT THE AUTHOR

Gina Lake is an astrologer with a Masters degree in Counseling Psychology. Since 1989, she has been in telepathic contact with Theodore, a nonphysical entity, and together they write about reincarnation, personal and planetary transformation, new age philosophy, consciousness, and healing.

Gina also works with Ascended Masters Sananda, Kuthumi, and Sanat Kumara, and is a frequent contributor to metaphysical magazines.

In addition to writing and counseling, she leads workshops on the ideas presented in her books, and is a noted speaker at metaphysical events.

DEDICATION

This is dedicated to all the Star People and seekers of planet Earth, who are giving birth to the Spirit within themselves and a new world.

Preface

The Confederation of Planets contacted me in July of 1996 and asked me to take down this message for them. They are preparing Earth for the changes ahead and for their own role in these changes. Soon, we will accept their presence, and this book is one way they are preparing the way. It may comfort you to know that I have my doubts concerning the existence of extraterrestrials. My experience of extraterrestrials thus far has only been telepathic. I hear them clearly in my head. I don't need to be in a deep trance or meditative state for this communication to take place. It is a simple process of listening to them speak to me and responding to them mentally with questions when they arise.

This process is not new to me, but communication with the Confederation of Planets is. Although I had heard of them before we met, I had no idea I would be chosen for this work. They told me that I agreed to work with them before I came into this life, but I was unaware of their presence before they contacted me. Apparently, there is a time for everything, and the time for this book and this message has arrived.

I still find it hard to believe that I am an author of two books on extraterrestrials. I had no interest in this subject before. It was with reluctance that I wrote my first book on this subject *The Extraterrestrial Vision*. I'm told that my disbelief and discriminating attitude is one reason I was chosen to write this. Despite my doubts, I find my path irresistible and ultimately rewarding. My interactions with beings in other dimensions have enriched my life immensely, and I hope that sharing this message with you will also enrich yours!

Peace to you!

Gina Lake, Novato, CA

Introduction

Who We Are

We of the Confederation of Planets, also known by other names such as the Federation, transmit this message to you, the people of Earth, through this author. We wish to greet you and tell you of our activities and plans. We wish to explain what we are doing here and how we can be of help to you. We come to you in service and in the Light of the Creator. Ours is a mission of service to you, for in serving you we also serve ourselves.

We are deeply grateful for this opportunity to communicate with you and for your willingness to read our message. We know how strange our words may sound to you, and yet in picking up this book, you are showing an openness to us and our message. We therefore write this message for those who are open enough to receive it. To you it is given. We are not here to convince you that we exist. You have only our words and their meaning by which to judge us. Many of you, though, will be able to feel us through these words, which act as carriers of our vibration. Those of you who resonate with this vibration will not find our message hard to believe. Those who do not will probably not even pick up this book.

The Confederation of Planets is made up of a group of extraterrestrials from various planets and dimensions in your quadrant of the universe. We are both self-selected and elected to our positions as overseers of this part of the universe, which includes your beloved planet. We are monitoring Earth out of love and a deep commitment to help her in her evolution and the evolution of all her life forms.

We have been with Earth in this capacity for eons, although those holding the various posts in the Federation have changed from time to time. For eons, we have watched as you developed your civilizations. We have watched and sometimes involved ourselves in your affairs. Early on, we watched as various others landed and populated your planet. Some of those who watch today were involved in the early experiments that created the human race. Some are even serving you today to pay off debts incurred then. So, you see, our motives for being here run deep.

We have learned much through our involvement and observations. We have studied the evolution of planets and the evolution of humanoids on other planets in order to understand how best to interact with you and guide you, for that is our task. We, too, once evolved in third density, the vibration in which you exist. Then, others watched over and guided us. Now it is time for us to repay this help through service to you. Someday you will play your part in the nurturing of planets and civilizations.

We come in peace, for primarily that is our hope for you. We live peaceably now among ourselves. We learned this through our many lifetimes of experience, just as you are learning this. We can help you become more peaceable by helping you to understand yourselves and by helping you prosper so that there is no need for conflict and strife. We have come to help usher in a Golden Age on your planet. This will not be for a while, but the seeds are being planted today for a much brighter tomorrow.

Truly, we have seeded the planet with what you need for this Golden Age. We have seeded the planet with many of a high level of development, who will help you solve your problems. These people have been called by many Star People, Starseeds, Lightworkers, and Wanderers. They are highly evolved beings who have chosen to incarnate on Earth now to bring forth new ideas and a new spirit. They are here to spread peace, love, and understanding by teaching and healing. They are also here to ease the suffering on Earth by introducing new inventions and advances in science and medicine, which will make it easier for your hearts to fill with love and gratitude for life and for each other.

We work closely with these Star People. They are like us, except they chose to take on a body to serve you. You see, there is much we cannot do because we are not of your dimension and much we are forbidden to do because of the Law of Noninterference. But Star People can do what we cannot. They are our hands, feet, and mouths, although they have free will and do not always choose to follow our suggestions. They are more able to hear us and intuit our guidance than the ordinary individual, however, and more responsive to us than most. We depend on them greatly.

There are many Star People among you and many more incarnating. Their numbers increase daily. But because of the density of Earth, many are unaware of their special powers or talents. They are not operating from the level of consciousness of which they are capable. Many are asleep. We are seeking to wake them up, for we need them to fulfill the tasks they chose to carry out before being born. They are a very important part of the plan to heal and transform Earth.

Your planet is in grave trouble, or we would not be watching you and involving ourselves in your affairs so closely. Nevertheless, the difficult juncture at which you have arrived has been anticipated for eons. This juncture is reached by most third-density planets at some point in their evolution. So, although these times are critical, they are not entirely unique. And because other planets have been successfully nurtured through a similar phase in their evolution, we are firm in our belief that Earth will make a successful transition too. How to do this is not just guesswork. We clearly understand what Earth is going through and how to assist her. You are not alone in this, and we are not alone in guiding you, for we, too, receive guidance from those more advanced than us.

We hail you, many of you Star People yourselves. We ask that you hear our message and let it resonate in your hearts. Hear us and awaken to who you are; awaken to your divine nature and to your chosen destiny! Arise and serve with us!

Blessings!

—The Confederation of Planets

ET CONTACT

Chapter 1

Who Is Here

Right now, thousands of extraterrestrials live in spaceships that circle Earth, and this is likely to continue even after some have settled on Earth. These spaceships are equipped with everything the occupants need to lead full and happy lives. The ships are like small cities—islands hovering in your skies—complete with waterfalls and other natural wonders. Many who live on these ships have never known life in any other environment and have no need or desire to relocate. They go about their daily lives just as you do, performing their tasks and carrying on their personal affairs. Not everyone is engaged in activities related to your world; some are simply support for those who are.

These ships are rarely seen because they exist in another dimension, although they can move in and out of your dimension. Although their higher vibration rate makes them invisible to you, you are not invisible to them. Those in higher dimensions can see and interact with those in lower ones, but the opposite is not true, except for those rare individuals who are clairvoyant.

On the ships, we have technology that helps us monitor the activities on Earth. So, when we say we can "see" you, we don't necessarily mean "see" in the same sense that your spirit guides and other nonphysical beings see you and know your every thought and feeling. Many extraterrestrials do have this clairvoyant capability, but most of the information-gathering that occurs on these ships uses technology.

There are many types of extraterrestrials and many types of ships in orbit around Earth. Some ships are populated with a variety of extraterrestrials who work together. Others carry only those of the same race. Their missions are many, their appearances are varied, and so are their methods; but they all have one thing in common: an interest in Earth. They are observing, conducting experiments, collecting data, and performing tests and other tasks, many too esoteric to explain. Most are here because your planet is going through a special time, a time of great transition and change, and they are interested in either observing or participating in it for one reason or another.

Besides the large craft that are their homes, they have small craft that make it possible to perform their functions. These are the craft most often sighted, for they dip in and out of your dimension as they do their work. Most of their work is carried out at night when they are least likely to be seen. For various reasons, they do not want to be seen. These small craft leave and enter via ports on the larger mother ship in which they dock when not engaged in their activities. They are very busy and can be seen in your skies nightly. They spend more time in some areas of your globe than others, but fly over even populated areas regularly.

The lives of the extraterrestrials on these ships are as real to them as yours is to you, even though you cannot see them. The fourth dimension, where most ships reside, is a physical reality, just like your third dimension. When you visit it in your sleep, as many of you do, it seems as real to you as the world you know when you are awake. That is why the contacts many of you have had seem so real. Most take place in this higher dimension while you sleep or in an altered state of consciousness. This dimension is very real when you are participating in it. When you have these experiences, most of you are in your astral bodies, which is the aspect of you that exists in fourth dimension, or fourth density as we will call it.

We use the word "density" as others may use the term "dimension." These words developed from two different systems for describing the same concept. "Plane" is another term that has often been used to express this concept. For our purposes, you can consider "density," "dimension," and "plane" as interchangeable terms, although our preference in this text will be to use the term "density."

Many densities, planes, or dimensions coexist simultaneously in space and operate independently. They are generally invisible to each other because each is of a different vibration, or frequency. Lower dimensions can become visible to higher dimensions; however, higher dimensions are invisible to lower dimensions except to some psychics.

The third density, which is where you reside, is a material plane. To those existing in fourth density and beyond, the fourth density is also a material plan, although it seems ethereal to those in third density. The densities beyond the fourth are not considered material planes and are populated by beings with no set form and no material needs. Humanoid evolution progresses through the various densities, starting with third density. You begin your evolution as physical beings in third density and evolve through the densities into nonphysical beings.

Some people do have an actual experience of contact in third density, but this is far more rare and is usually reserved for Star People and Walk-ins. The reason for this is that most people cannot handle such direct contact; it is too shocking. Generally we do not interfere so blatantly in your lives; however, some people have made soul agreements with us that give us permission for this kind of contact. Usually it is so that these people can help us with our mission.

There are higher density beings on these ships, too, beings of fifth and sixth and sometimes even seventh density, who are assisting. Usually they are in positions of leadership. The Confederation of Planets, for instance, is made up of a variety of beings from many different dimensions and planets. These beings are organized hierarchically not only because it is more efficient but because it also enables us to be more effective. This hierarchy is accepted by us, and we do not have power struggles or conflicts over position and tasks. We perform our work in an orderly, peaceful, and dedicated manner.

What Is an Extraterrestrial?

Any being that is not from Earth or living on Earth at this time could be called an extraterrestrial. This includes higher dimensional beings, whom you might call spirit guides, Ascended Masters, and angels. This does not include spirits, or those still on the wheel of reincarnation, who await rebirth on your planet, but it might include other third-density beings who are not from Earth. By "from Earth," we mean beings who spend a majority of their evolution in third density on Earth rather than on some other planet. This is complicated, however, by the fact that nearly everyone spends some of their physical lifetimes on planets other than the one on which they are primarily evolving.

You are used to thinking of extraterrestrials as little green men from outer space, that is, beings with physical bodies like yours. Well, some do have physical bodies, some like yours, and some very different. There are all manner of humanoids and non-humanoids in physical bodies in this vast universe. There are even humanoid extraterrestrials that look just like you because they are related to you genetically. Those interacting with you in physical or quasi-physical bodies are either third or fourth-density extraterrestrials, who are indeed physical and not capable of changing their appearance as a general rule. (OR FREQUENCY)

But many of those interacting with your planet now are not physical. They are fifth-density and beyond. Most of the work these nonphysical beings do is of a nonphysical nature. They usually don't take on bodies and affect the physical plane that way. Many of the higher dimensional beings working with Earth do not and will not choose to appear physically to people, although they may appear in a vision or dream if that serves their purpose.

However, there are some higher dimensional beings who do on occasion materialize a body that appears physical to you. If they do, it will be an appearance of their choosing, one that fits their purposes. They are not identified with one way of appearing, although some do have a favorite persona or personas. Others don't materialize a body but appear to people as Light Beings—large ovals of light with vague features that identify them as humanoid.

Seeing a Light Being is likely to be a profound experience for most of you, while seeing a third or fourth-density extraterrestrial is more likely to be frightening, not only because of their realness but because some of them are scary to look at from your point of view, or they may not operate in the Light, that is, in service to you. However, there are even some fifth-density beings who are not operating in the Light. If you encounter one of them, you will know by the confusion they invoke. With Light Beings, you feel love, peace, joy, and a sense that something profound is happening to you.

Two Varieties of Extraterrestrial

Both positive and negative extraterrestrials are involved with your world today. Positive extraterrestrials are here to serve you, while negative extraterrestrials are here to serve themselves and feel no duty to serve others. Negative extraterrestrials feel that serving others is a position of weakness. Eventually, all those on the negative path do become positive, but some stay on the path of service-to-self for eons before making this switch. They serve the role of villain in those realities that evolve through the duality of good and evil. In that way, they do serve the Creator, although they don't see it as that of course. They serve the Creator's plan because evolution progresses by running into obstacles and overcoming them. They provide the grist for the mill of evolution. + pain and suffering

We of the Confederation of Planets are positively-oriented, but we do allow those of a negative orientation to interact with you since it is part of your evolution to experience them. We are pledged to help you but not to interfere with your evolution. Thus, service to you does not mean giving you everything you want or protecting you from your challenges or from experiencing the consequences of your choices. We are here to influence you positively, but we honor your choice to ignore our nudges and suggestions.

Our attitude toward those oriented toward service to self is similar: we encourage however we can, but we do not interfere with their choices, which would be interfering with their right to learn in the manner they choose. We respect their right, as well as yours, to choose a course of action. However, when these choices have potentially far-reaching consequences to the well-being of the planet and its life forms, we feel a duty to intervene. This is why we are involved more directly with your planet now.

Your planet is dangerously close to destruction, both through nuclear mishaps and environmental degradation. The loss of the planet Earth or its ability to sustain life, especially sentient life, would be a loss for the universe and would have particularly far-reaching consequences for this quadrant of the universe, which we oversee. We therefore feel a duty to do what we can to prevent such damage from occurring and to help you correct the damage you have caused to your environment, if you are willing. We are here to teach you of peace, and we are here to help you live more in harmony with the blue-green jewel that sustains you.

The negative extraterrestrials are here to perpetuate your difficulties. They are drawn here for the same reason the positives are: they see that you are vulnerable and on the verge of cataclysm. The difference between the positives and the negatives is that the negatives hope to benefit from your misfortune and hope you continue on your current course. They try to influence you in many of the same ways as the positives do: through dreams, intuition, feelings, telepaths, psychics, and Star People and Walk-ins.

Star People and Walk-ins

Some people are of extraterrestrial origin, although they are in a human body. There are two ways these higher dimensional beings come to Earth: through the usual way, reincarnation, or by walking into a body. Those who reincarnate are called Star People, Wanderers, or Starseeds; those who walk into a body as the soul is leaving it are called Walk-ins. This soul-exchange happens without the body dying. Star People and Walk-ins can be of either orientation—positive or negative—although most from fifth density and beyond are positive.

Most negative Star People and Walk-ins are from fourth density. Negative Star People and Walk-ins from fifth or sixth density are rare but especially capable of causing problems on Earth because of their high intelligence and special gifts. These are the unscrupulous individuals who often gain positions of power and abuse these positions. They are highly manipulative and, by your standards, immoral. They do have a moral code of sorts, which is, "If you can get away with it, do it." They are self-serving and will stop at nothing to achieve their goals. There is no limit to the kind of suffering they are willing to cause to get what they want. They are not averse to causing suffering because they believe that victims deserve their lot. Because they will do anything to avoid being victims themselves, they believe others should too.

We do not wish to dwell on these individuals. They are not news to you; you've lived with them for some time. It is important, though, to realize that there are such individuals in your world, since most people find their behavior incomprehensible and therefore often give them the benefit of the doubt and then fall prey to them. There are more negative Star People and Walk-ins on Earth today than ever before, but there are also more positive ones. The negatives are not only outnumbered by the positives but also outranked, since the positives are not only from fourth and fifth density but beyond.

Star People and Walk-ins are playing crucial roles in the transformation of your world. They are often in key positions because of their talents and intelligence, and are usually found on the forefront of change. They have easy access to information from other realms and act as conduits for new ideas, often without knowing it. Many don't know who they are, but that doesn't necessarily stop them from carrying out their mission. Nevertheless, many Star People are still asleep and are not applying themselves to their life purpose.

Many of the Star People will regain full consciousness and knowledge of who they are. When this happens, the effect will be obvious in your world. Not only will they function in their work at very high levels, but they will bring in a degree of compassion and unconditional love that few have ever encountered. Christ is surely coming again, this time in the form of many Christed individuals. And the more Christed individuals who walk the planet, the easier it will be for others to come into their divine consciousness.

The current period is perhaps the most difficult because just a few of you have awakened to this level and are looking around and saying, "Hey, where is everyone? I could use some help!" or, "What's happening to me?" Many of you are struggling with your new identities and wondering if you are crazy for seeing things the way you do. We are here to say that you are not crazy.

Many of you have always known there was more to life than meets the eye. Many of you have had paranormal experiences, near-death experiences, or traumas, which have caused you to question the meaning of life. You have often chosen challenging lifetimes on Earth and may be living a very challenging one now. Most of you who are Star People have spent hundreds of lifetimes on Earth learning about being human in preparation for what you are doing today. Coming to Earth to serve during these times was a big commitment on your part. You are valiant souls indeed to undergo the rigorous training this commitment requires and to reexperience the suffering of third density. We deeply respect this choice and do all we can to make your work and your awakening easier.

One of the marks of a Star Person and Walk-in is a sense of loneliness and longing for home, and the memories of a more beautiful and loving place that is your real home. You gave up that place to come here and serve. Sometimes you need reassurance that it was worth it, that you are succeeding, that you will accomplish what you set out to do. So we work closely with you both when you are asleep and when you are awake. While you are awake, we work with you intuitively, as your guides and other nonphysical beings. While you are asleep, we provide healing, apply consciousness-raising technologies, take you on visits to other planets and dimensions, teach you, love you, and help you in any other way we can. We are your family, and many of you feel this when you are able to remember our contacts with you.

Many Star People and Walk-ins are telepaths for us and other extraterrestrials. Telepaths perform a very important service in being direct mouthpieces. One word of warning, though: even good telepaths don't always communicate with high level beings or do so consistently. There are many types of beings out there clamoring for a vehicle to speak through. Many have very specific agendas that are not necessarily in Earth's best interest right now, although they may believe they are here in service. It is important for you to understand that not only are there negative extraterrestrials who are trying to scare, confuse, and otherwise throw a monkey wrench into things, but there are well-meaning extraterrestrials who just don't understand humanity and its problems well enough to be helpful. There are also those who, although well-meaning, seek to impose their own perspective on you.

Those of you who are Star People feel a sense of alienation from life on Earth. You wonder how people can behave as they do; how they can possibly engage in war, for example. Humanity looks quite insane to you. Some of you seek comfort in drugs that numb your feelings and give you a glimpse of the peace of the dimensions you know so well. Some of you end up in mental institutions or suffer in other ways, finding it so difficult to cope with life here. You do your best but until you awaken, there is a restless gnawing inside you, nudging you to know who you are.

Many of you pore over metaphysical tomes, memorize incantations, become adept at yoga postures and meditative techniques—all in search of yourself. Most of you have been searching all your lives. It can take that long to awaken! There is so much to overcome. And it takes courage to look deeply at life, to be different, to ask questions. We honor you and your journey so much! We hold you deeply in our hearts. You are part of us and we are here to bring you home. But first you have work to do.

Fourth-density Beings

Those in fourth density, which is the next dimension beyond yours, are physical. Like you they have physical bodies that require sleep and refreshment. However, their physical needs vary widely, and some don't appear to need food as you know it. Nevertheless, they do depend on the environment for their survival and therefore appreciate its importance. Some are here in part to teach you the importance of proper stewardship of your environment.

They are vulnerable to the same problems of having a physical body that beset you: accidents, sickness, disease, and aging, but their technological and intellectual advancements make it much easier for them to deal with or avoid these things. In addition, those who are also spiritually advanced know how not to create illness. For these reasons, there is little physical suffering in fourth density. Their energy is free to focus on tasks other than survival and physical well-being.

Fourth-density beings move quite easily in and out of your dimension by stepping down their vibration and dematerializing and then rematerializing their bodies. They appear to you as they appear in fourth density. However, most cannot take on other forms, as can beings beyond fourth density. When they are in your dimension, they appear solid and real. Many who have seen an extraterrestrial in this dimension are unable to remember it because it is so shocking. Most lose consciousness or simply block out the experience, and all that remains is a sense of uneasiness and "missing time" that cannot be accounted for.

Most abductions and other contacts with extraterrestrials occur in fourth density, sometimes when the person is in the astral body and sometimes through the use of technology that translates the person's third-density body into fourth density. It is then translated back again when the contact is over. This is the reverse of how the extraterrestrials contact you in third density. Usually, however, it is easier to contact you while you are in your astral body, so that is the most common method.

Examples of fourth-density extraterrestrials are those from Zeta Reticuli, including the Greys, the Pleiadians, the Reptilians, some Orions, and some negative Sirians, although there are many other fourth-density races involved with you now. Many extraterrestrial races have members in a number of different densities, not just one. The Pleiadians, for instance, have counterparts in fifth density and beyond, but the majority of those contacting you now are fourth density. Those in higher densities are beings who have simply been around longer and now work to guide those in lower densities. Most of those we call higher density Pleiadians, for example, are drawn to guiding lower density Pleiadians because of strong ties formed with the Pleiades during their earlier evolution.

The fourth density is an etheric world to you, but a very real one to fourth-density beings. When you travel there in your astral body, it seems real to you too but, on your return, any memories seem unreal—like a dream. This is why you often doubt the reality of your experiences there: fourth density does not seem real from the standpoint of third density. Fourth-density beings feel the same way about third density after returning from it—the experience feels dreamlike. The difference is that they understand that other dimensions are real even if they don't seem real.

Fifth-density Beings

Just as fourth density seems etheric to third-density beings, fifth density seems etheric to fourth-density beings. They can travel there, just as you can travel to a higher density while you sleep. And like fourth density to you, fifth density to them seems dreamlike and difficult to recall. They are refreshed by traveling there, which they do periodically in a state similar to sleep. There they receive guidance, healing, and teachings. It serves the same purpose that fourth density serves for you.

Some of you also travel to fifth density when you sleep. It is where you are most likely to meet with spirit guides and others significant to your spiritual growth and development. When these experiences are recalled, they feel heavenly and profound.

Fifth-density beings no longer have physical bodies. They are beyond the physical lessons, but serve the physical world through teaching and guiding. This is the realm of spirit guides, who are assigned to help people in third density. The guides do this primarily through working with people's unconscious minds while they are asleep and awake, and by sending messages to them intuitively and through their feelings. In these ways, they influence people's thoughts, feelings, and choices.

Fifth-density beings are wholly devoted to service and have no personal agenda. The exceptions to this are those who are still on the service-to-self path, but they are in the minority. This doesn't mean, however, that fifth-density beings are perfect, even if their motives are pure. They are still learning, and they learn by making mistakes just as you do. If free will did not exist, their job would be infinitely easier. But since people are highly unpredictable, fifth-density beings sometimes fail to judge a situation correctly. It is good to keep this in mind if you ever meet a fifth-density being because you will be overcome with love, and it will be easy to assume that that being is perfect.

Sixth-density Beings

These are very high level beings who act primarily as teachers and guides for those in fourth and fifth densities. They generally are not involved with third-density individuals except for some telepaths and Star People. The love they exude is so intense that, as with angels, an encounter with one sends most people into tears of rapture. This is one reason they do not take on physical bodies and interact with third density. When they interact with people telepathically, they use only a fraction of their energy. Another reason they don't take on physical bodies is that their energy is better used in other ways. Most are involved in many tasks in many dimensions and locations at once and leave lower-density tasks to others.

Who's Been Here

Extraterrestrial visitors are nothing new. They have been visiting Earth since before humanity existed and were, in fact, responsible for its seeding. Many different groups of extraterrestrials have overseen the activities of Earth at different times. Certain groups have always been here and will remain. Primary among those are the Pleiadians, who are very deeply connected to this planet and to the human race. Neither have the Sirians or the Orions ever left your planet.

The Greys, on the other hand, are relatively new to your planet. Some extraterrestrials, like the Greys, have learned of you only recently because of the fix you are in. And some just happened to run

across you in the course of their travels. Except for the Greys, these newer arrivals are less important than the long-standing groups, many of whom have karmic ties with you. Those who have been involved with you throughout your history, especially with the creation of the human race, are tied to you by bonds of both love and karma.

It may not please you to know that many of the extraterrestrials involved with you now are trying to make amends for ancient mistakes. We tell you this because it is crucial to understand that even now some of these groups are creating *more* karma for themselves rather than less. These well-meaning groups will confuse you the most. They will seem loving and well-intentioned, and they are; but that doesn't mean they are doing what is best for you. If you allow yourself to be victimized by them again, it will be your lesson. We warn you of this to help you avoid making the same mistake again.

We have given them advice to interact with you differently, and some are heeding it and backing off. There are always the few, though, who are so certain of themselves and unaware of their own motivations that they will not take any advice. In some cases, some service-to-self beings have infiltrated their ranks and are influencing them without their knowing it. It may be difficult to see how this can happen to beings who are this evolved, but evolved service-to-self beings are very clever. From time to time, they trick even fourth- and fifth-density beings, but usually not for long.

You are on an evolutionary journey. So are they. You are learning, and they are learning through their interactions with you. There is no blame in this; it is the way of evolution. Making mistakes and correcting them is how we all evolve. No one is wrong or bad for making mistakes. Mistakes happen, and sooner or later those same mistakes are no longer made. No one is exempt from this process! We hope you can have as much compassion for the extraterrestrials who have harmed you in the past as you have for yourselves and your own mistakes. We are all growing. And we are all growing toward the Light!

Chapter 2

What We Are Doing Here

Healing and Teaching

We are here to uplift your race, and most of our work involves healing and teaching. Helping in these ways is not considered interfering in a race's development but just a natural part of evolution. It is our joy to do what we can to help you evolve as individuals and as a race.

Part of our healing work relates to working on the planet's grid system to help the planet adjust to the changes in its own vibration and to help the human race open up to galactic citizenship. In order to participate effectively as citizens of the galaxy, human consciousness must be raised. This must occur in any event, since the human race is dangerously close to destroying itself and much of the planet along with it. Adjusting the grid system will allow a higher frequency of energy to enter the planet's field and be absorbed by those capable of receiving it. So, we not only send a higher vibration of energy to the planet, but make changes within the planet itself to make it possible for more people to receive and integrate this energy.

Healing is critical if your bodies are to integrate the higher vibration that Earth is moving into. This is why healing is such an important part of our work here. You are all wounded simply because you live in societies that do not foster healthy development. So, even those of you who are very advanced souls find it hard to break free from the conditioning you received while growing up in this world. Being human isn't easy for anyone!

Star People may have special talents and perceptions that often lie buried by the fear and pain that fill your world. They need ways just like everyone else to get beyond the fear and to heal the pain so that they can reach their full potential and fulfill their life mission and purpose.

Our work requires individuals incarnated on Earth who can hold the higher vibration while others adjust to it. These individuals are usually Star People, but anyone who can hold and transmit a certain vibration may be used for this. These individuals are entry points and anchors for the higher vibration. Star People and Walk-ins are especially able to express and maintain higher levels of consciousness although not all are currently doing that.

Helping you adjust to and integrate a higher vibration of energy is an important part of helping you create a new world. This is happening naturally and affects all of humanity, which in turn must adjust to this new vibration. Bringing this energy in through people who then transmit it to others just by their presence exposes humanity to this energy relatively easily because people will take in only as much energy as they can. With increasing exposure, their ability to receive and integrate the new vibration into their own energy bodies increases. This higher energy brings about healing in their emotional bodies, as it clears away blocks. This is one way we are working to heal the planet.

Another way we bring healing to the planet is by inspiring people, especially Star People and Walk-ins, to develop new methods of healing. Many of you who are Star People were already healers where you came from and have special knowledge that you share. You are pioneers in the healing field on this planet, but your techniques are not new or even new to you. You pioneers are being reminded of this knowledge through your dreams, intuition, visions, and other psychic impressions. Many of you practice your healing arts in other dimensions while you sleep.

Our teaching work includes sending information through your unconscious minds to help you advance as a race spiritually, emotionally, intellectually, and physically. We send you useful scientific and mathematical information, new healing methods, ideas for improving educational and political systems, music and poetry that uplifts, and science fiction that opens minds and prepares you for what's ahead.

Many, many people are involved in receiving this information and inspiration from us, and although most are unaware of the source of their ideas, many do act on them. We work especially closely with Star People and Walk-ins, many of whom are teachers, inventors, artists, musicians, writers, scientists, telepaths, psychics, and others with special gifts and a special vision for Earth's future.

Many of us work directly with telepaths and psychics to heal and teach. They allow us to work through them to express our particular perspective or deliver energy to the planet. Because being here ourselves would be too much of an interference in your evolution at this time, we count on being able to influence the physical plane through those who are physical and freely choose to act as vehicles for us. They act as focal points for our energy and our ideas, and without them we would not reach as many people or reach them as effectively. Of course, beings like ourselves have always influenced Earth in this way, but there are many more vehicles for us now than ever before. That is all part of the plan for these times.

These telepaths and psychics, however, are at various levels of development and some have not fully unfolded their talent. Not all of them operate as effectively as they might. Those who are undeveloped or unclear can hinder us rather than help if they fall prey to negative forces who may use them to confuse, trick, or deceive people. So, although telepaths and psychics are some of our best tools and we rely on them heavily, they are also a potential liability.

Because many of the Star People and Walk-ins have psychic abilities, we use them as instruments to perform certain activities important to our goals for Earth. They are the hands, feet, and mouths of those guiding the planet from higher densities. Needless to say, they are very important to the extraterrestrial mission. However, most have not yet awakened to who they are or regained their original level of consciousness. This puts a greater burden on the ones who are awake, who feel stressed by both the lack of support in the world for their point of view and the slowness of change.

We have consciousness-expanding technologies that we employ on the unawakened Star People while they sleep. This has some effect, and soon large numbers of them should awaken to the point where they can take advantage of new technologies such as brain synchronizers, biofeedback machines, and subliminals, along with the many new healing methods, particularly those using color and sound.

Many Star People are involved in businesses that promote and employ these technologies. Their main thrust is to awaken and heal as many other Star People as possible, who can then awaken and heal others and act as entry points for the higher vibration energy. The consciousness of the human race is being raised now through many of these very practical and tangible means. It is not entirely esoteric!

Guiding

Many of us serve as guides for people on Earth, especially for Star People and Walk-ins. We guide people through their intuition, dreams, feelings, and other unconscious means, and more directly through telepaths and psychics.

Being a good guide is an art, as is being a good therapist. A guide has to know what to say, how much to say, and how and when to say it. After all, a guide's job is to influence the attitudes, beliefs, thoughts, feelings, and direction of the person being guided, all without interfering with that person's free will. Guides receive extensive training in how to do this. Their job is especially challenging when it comes to introducing information that is difficult for the person to believe or accept. The guide will have to gradually accustom the person to these ideas.

Therefore, the persona that the guide takes on is important and is chosen carefully. Most guides are beyond personality, and must draw on their past life experiences to interact with the person they are guiding. Since most people aren't able to see or speak to their guides directly, for many, the guide's persona is not important. But when working with telepaths and psychics, it is very important.

To avoid interfering with the person's free will and the natural unfolding that results from it, one of the primary rules in guiding someone is to avoid introducing unsolicited information that he or she has not asked for or is not ready for. This can be difficult to determine, so the person is carefully studied. Most guides have studied the person they guide for many lifetimes before taking on this task.

There are many ways to a goal, and guides honor this. They work with a person's choices, gently nudging him or her along courses that are more favorable to the soul's plan, mostly through the unconscious and dreams.

Highly evolved beings do not interfere with your choices, but they do make themselves available for advice. You must open up the topic for discussion, though; they will not raise an issue about which you have no thoughts. They will interact with you more freely on a subject that is already in your mind. And although they might lead you to a new line of inquiry and new subject areas in their talks with you, they will wait for you to pursue that new topic with questions. Sometimes you unconsciously don't want to know something, and your guides respect that.

Something else that is useful to understand about guides and others who work with human beings and the planet is that they do it also for their own growth and evolution. Their service to you serves their own evolution. They learn through their interactions with you. They don't have all the answers nor do they always know the best way to go about accomplishing their goals. They try this and that until it works. In this way, they are much like you. They are learning, too, but in different areas. They have already learned what you are learning, but that doesn't mean that they know how to teach you what they know. They learn by trial and error, just as you do. Guides are not all-knowing, nor are the extraterrestrials you will meet in person. In fact, many of the extraterrestrials involved with Earth are only fourth-density, the dimension into which Earth and humanity are moving.

Abductions

The word "abduction" does not accurately describe the activities of the positive extraterrestrials, although it fits the activities of the negatives. For lack of a better word, we will use it since the event itself, when remembered, often feels like an abduction, whether it is by positive or negative extraterrestrials.

Most abductions are not remembered. They do not occur in normal waking consciousness, but in a dreamlike state where memories are not readily accessible. Furthermore, both positive and negative extraterrestrials can prevent memories from being carried into waking consciousness, although these methods are not always totally successful.

Memories are suppressed by positive extraterrestrials to avoid shocking the individual. Even though those abducted by the positive extraterrestrials have made soul-agreements to help us, they do not remember these agreements and often feel traumatized by such an extraordinary experience. Even the most stalwart of souls has difficulty accepting and integrating this experience when remembered.

Aside from the shock of these experiences, abductions by the positives do not inflict pain, nor are they intended to induce fear. The positives go to some lengths to make sure that the abductee is comfortable and that the experience is as painless as possible. They have the technology that makes it possible to eliminate the memory of the pain, if not the pain itself, which is effectively the same as eliminating the pain as long as the memory is not somehow uncovered.

Abductions by negative extraterrestrials, on the other hand, occur without a soul-agreement, but are at least tolerated by the soul because it can serve the person's growth. This tolerance is quite different from the soul-agreement most Star People have made with us prior to incarnation.

As much as we appreciate what abductees have done to further the understanding of the extraterrestrial situation by uncovering their memories, we do discourage attempts to recover these memories. They were covered for a purpose and are best left alone. We do appreciate, however, that these buried memories influence people's lives negatively through anxiety, nightmares, and other means by which the unconscious seeks expression, and we understand your need to relieve yourself of these symptoms. We would have to say, however, that in this case the cure is worse than the symptoms, and repressed memories are better left alone. There will come a time when it is no longer necessary for our presence to be hidden from you, but for now there are good reasons for this. Unlike the UFO flybys, the purpose of the abductions is not to prepare people for the truth of our existence but to provide us with the information, genetics, and biological material we need to fulfill our mission.

The negatives abduct those who have a tendency to play the victim, and this is allowed in order to teach the person not to be a victim. This is not blaming the victim but rather simply how spiritual laws work in third density. Victims attract negativity, which perpetuates the situation. Everyone incarnating on Earth has experienced or will experience victimization, since it is part of your evolution here. You move beyond it by standing up to your oppressors and refusing to be victimized.

Some argue that standing up to the negative extraterrestrials doesn't work because they paralyze their victims, making it impossible to fight back. Nevertheless, it is possible to change one's psychological orientation so as not to attract this experience in the first place—and that is the lesson. This is done by not playing the victim and by maintaining a strong sense of identity. Those who are assertive, know what they want, and refuse to be manipulated by others are not chosen by the negative extraterrestrials.

Abductions have been occurring since the 1940s, but are now more frequent. More and more people are spending time in their sleep state assisting extraterrestrials. This is not only because our need for this assistance is now greater but because there are more people incarnate now than ever before who have made agreements to help.

The number of Star People on your planet has increased tremendously in the last two decades. There are more children born today who are Star People than ever before. Even those who are very young are helping, for age does not prohibit the astral body (the aspect we work with most often) from interacting with us. It is not necessary for Star People to be adult to work with us while they sleep. The presence of more Star People on Earth is evidence of our imminent arrival upon your world.

Hybridization

The abductees, who are mostly Star People, help us in a number of ways, one of which is to donate sperm and ova to produce a hybrid race. This is not an attempt to improve or save the human species as much as to save the race of extraterrestrials involved in the exchange. This type of crossbreeding is as commonplace throughout the universe as agricultural crossbreeding on Earth.

Genetic engineering is simply part of the learning and evolution of species, sometimes failing miserably and other times leading to great advances. While not considered unethical by those involved, how they go about it or what they do with the results may be unethical. Sentient beings have a curiosity and drive to explore, learn, and discover, which they express in these experiments. They learn important lessons and advance their knowledge through these experiments.

Most of the hybrids will be relocated elsewhere; only a fraction will remain on Earth. But this small group of hybrids will be responsible for much transformation in your world. This is why the abductees have volunteered to help—they know they are greatly serving the future Earth. The hybrids have an extremely high intelligence quotient by human standards, and their intellectual gifts will advance science and society. Many will be great leaders and on the forefront of change in the 21st century. They are spiritually as well as intellectually developed, which is essential to being positive forces for change. They know all too well what happens when a race's spiritual development lags behind its intellectual development. Very few of the hybrids are adults yet, so it will be a few years before you experience their impact in your world.

These hybrids are primarily a cross between Earthlings and Zeta Reticulans, those short, spindly beings with the large head and eyes you've seen so much of lately in drawings done by abductees. (The negatively-oriented extraterrestrials from Zeta Reticuli are often called Greys.) While other genetic experiments involve other extraterrestrial races, this particular experiment is critical to the survival of the Zeta Reticulans and it is the most wide-reaching, involving hundreds of humans each night across the globe. Although it has taken quite a while to reach this point, it is also one of the more successful hybridization projects.

Other abductees help out as needed in the many other tasks necessary to the hybridization project. For instance, the hybrid children need to learn from and be nurtured by humans in order to grow up healthy. Because they are part human, they have the same emotional needs as human children. But the Zetas do not know how to nurture human children, and learn this from the Star People and other humans who have agreed to help. Thus, many Star People spend their nights in nurseries and play areas nurturing and caring for these children.

Long ago, the Zetas programmed emotions out of their genetic structure but now see the value of emotions and study yours as part of recreating them within themselves. Star People and other abductees also have volunteered to allow the Zetas to study their emotions, both behaviorally and chemically. So, through hybridization, the Zetas are attempting to regain the emotions they lost through their own misguided genetic engineering. They hope to learn how to interact with you in ways that are less intimidating because the positive Zeta Reticulans really do not mean to frighten you. They are learning empathy for your fear as well as how to calm you. They are studying the chemistry of your emotions for both these reasons.

Many of the Star People and other abductees who are helping with reinstating the emotions of the Zeta Reticulans are reincarnated Zetas themselves! This is one way the Zetas are helping themselves. They are allowed to reincarnate on Earth because they give you a gift in exchange—the wisdom they have gained from traveling down the wrong path, a path that your race could easily follow. They are here not only to help themselves but to help you by warning you not to make their mistakes. They are not concerned that you will do away with your emotions like they did, but they are concerned that you will do away with your planet. They, too, had nuclear weapons and misused them, and are deeply concerned that you will do the same. They are balancing their karma around this by bringing you this message of peace.

The Zetas look strange to you, but they didn't always look that way. Long ago they were similar to you, since they are from the same galactic family and they were seeded by the same ancestors as you. They mutated and changed their appearance primarily through genetic manipulation to help them adapt to their planet contaminated by nuclear war. The changes they are making in their race as a result of their encounters with you will restore a more human form, with all that goes with that. They look forward to this, but it will take eons before the changes are complete. Their life span, however, is at least twice yours, so the change will take fewer generations. Besides, they have no other alternative.

Peace

Star People and other abductees are involved in many other experiments and tasks related to the Zeta's transformation. Some of them volunteered simply out of service: there was a need, so certain individuals applied to fill that need. Others volunteered because it would balance some karma they had with the Zetas or because of a similar situation on another planet that was destroyed by nuclear weapons. By helping the Zetas, who destroyed their planet and most of their race and are now atoning for that, these helpers support the message of peace the Zetas have for your world and others. Therefore, any being interested in promoting peace and ending war might volunteer to help the Zetas. The Zeta's mission is essentially one of peace, and many beings throughout the universe want to serve peace.

One aspect of this peace mission involves gathering samples of your flora and fauna to preserve in case your planet is destroyed. The Star People involved in this aspect of the Zeta's mission are primarily scientists and come from planets that are very advanced scientifically, particularly in the areas of molecular biology and genetics. With the Zetas and other interested extraterrestrials, they study the life forms on your world because this yields useful information for those on other worlds who are creating life. In exchange for the privilege of studying your life forms, these extraterrestrials and the Star People working with them serve by preserving your life forms for you in the event of a catastrophe. At a certain point in your evolution you, too, will become such creator-gods.

Ecology

Some of the abductees have volunteered for another aspect of our mission: to help the planet environmentally. Here again, many of those volunteers are balancing karma incurred from harming the environment of some other planet. They are learning the importance of taking care of one's home, and they hope to help you learn that lesson as well as help you save your planet. They work closely with a group of various races to monitor the environment and create strategies for rectifying the environmental damage on your planet once you ask for our help on this matter. Help of this kind will be offered to you soon after our formal introduction. These Star People help us collect and analyze data gathered from around the planet, often from the UFOs seen nightly.

We are involved in more than just monitoring the environment and analyzing data; we also correct problems, not so much on a physical level as an etheric one, by repairing the planetary life-support energy grids and matrices. Much of our work involves energy and the manipulation of Earth's energy field. Just as healers today help people heal by altering the flow of energy in their energy fields, or auras, we help the planet heal by working with the flow of its chi, or energy: "As above, so below." But our advanced understanding and technology can only do so much. These grids can be damaged beyond our ability to repair them or repair them in time, which is where the concern lies. Yes, we can help you, but you must also change your ways or our help will be insufficient.

Negative Extraterrestrials

The negative extraterrestrials are involved in some of the same tasks as the positives, but for different reasons. One negative race, the Greys (or negatively-oriented Zeta Reticulans), also are trying to save their race through hybridization with your species. The Greys are finding, however, that unlike the positive extraterrestrials, their skills are not advanced enough. On the whole, negative extraterrestrials are not as technologically developed as the positives because their negative orientation interferes with their evolution.

The negatives do not see any use for emotions. Unlike the positives, the negatives do not want to restore their emotions; that is not why they are involved in hybridization. They study your emotions not to understand you or develop empathy for you, but to control you. They hope to colonize your planet and rule you, and they need to know how to instill fear in you so that you will submit to them.

The negatives, too, are studying your environment because they need to understand it to live here, something they currently cannot do without strain. As it is, their life span is reduced considerably by being here, and it will take many hundreds of years for them to adapt enough to be comfortable here for long periods. Their plan is to live between their spaceships and the planet during the time it takes to adapt. This is also the approach that most others who settle on your world will have to adopt.

The different negative groups involved with Earth now have various agendas, many of which conflict with each other, and this will make it difficult for any one group to gain supremacy. The negatives are at cross purposes not only with the positives but with other negative groups as well. Nevertheless, some negative groups do work together on some projects, although they do not cooperate well with each other. The negatives are learning cooperation through the consequences of not cooperating. While to most of you this seems like a simple lesson, negative extraterrestrials and negative human beings have a different outlook on life.

In general, the negative extraterrestrials are busy planting seeds of discontent in the unconscious minds of a broad spectrum of people and provoking negative thoughts and feelings that people believe are their own. Mostly, they hope to undermine and inhibit the Star People, who they see as their greatest threat among those who are incarnate. It is no wonder many Star People feel inundated with negativity at times. The negatives are especially able to influence those who are under stress or struggling with negative emotions. That is why it is so important for you to stay healthy, both physically and emotionally.

The negatives influence your actions and choices in the same way the positives do, but it is when they get hold of people in positions of power, such as within the media, that they can cause the most trouble. Because they try to influence those who have the most power in your world, anyone in such a position needs to be on guard and live impeccably. Unfortunately, there are many negatively-oriented human beings in positions of power who serve the goals of the negative extraterrestrials.

You are influenced by outside forces more than you realize. You may be aware of the power of media to influence you, but are you aware of the forces that influence the media and your own thoughts and feelings?

Awareness will help you take control of your thoughts and feelings so that you do not become targets for the negatives. Just being aware that negative influences exist and that any negativity you experience may not belong to you can help you stay in control and fend it off. We tell you this not to scare you but to empower you through knowledge. This knowledge disempowers those negative forces who use ignorance to gain a stronghold in people's minds.

Chapter 3

How Star People Are Helping Us

Many of you reading this are Star People and we want to dedicate a chapter to you so that you can better recognize yourselves. We don't mean to limit this to Star People; Walk-ins and many other advanced souls are also included in the following descriptions. But for simplicity, we will refer to all those helping us as Star People. These descriptions also reveal the direction in which we are encouraging your world to move. So they serve two purposes: they alert you to your mission and describe our mission to you.

Star People in Education

Because education is so important to the preservation and transformation of society, many Star People are here to effect change through your educational system. Many of those in this field are innovators, finding new and better ways to teach children and heal the system. Right now you possess all the new ideas and understanding you need to transform your educational system. The ideas are there, waiting to be implemented on a larger scale. The problem is that the system resists change and will only try out new programs when it becomes desperate.

Your educational system is breaking down because it can't keep pace with the times and the many changes that the information age has brought. Information is becoming obsolete before it is even grasped. Much of what is taught in your schools is not relevant to the world of the future.

An even more serious concern is the fact that you are not teaching people to think and be resourceful. Moreover, you are not teaching them to love learning. Your attempts to train and program children with information kills their innate curiosity and discourages them from pursuing learning in their own way. They need not only the freedom to learn in their own way but also the freedom to learn about the things that interest them.

The innovators, many of whom are Star People, respect the innate sensibilities of children. They trust that everyone has unique gifts and they know that unfolding those gifts requires a unique approach. They trust that people intrinsically know what is right for them and will pursue it if allowed to. And they trust that doing this will bring fulfillment, happiness, and productivity.

The innovators believe not only in free thought but the freedom to choose and to act according to one's hopes, dreams, feelings, intuition, and drives. They encourage not only free thinking but freedom. Many of them have lived on other planets or in earlier times on Earth where this freedom was squelched, and they are here now to fight for this freedom through the educational system. They know that a free society must create freethinkers, and that this is done through the educational system. If you doubt this, just look at the educational systems of societies that are not free. Notice how they train their youth to perpetuate their enslavement.

Star People in all walks of life are freethinkers. They often hold opinions or viewpoints that differ from the mainstream. That is because most of them are here to change the status quo, not support it. One of the things that needs changing in your world is the belief that people cannot be trusted, that they are basically sinful and bad. This belief is at the base of your oppression. How can you allow people to be free if you don't trust them?

Star People understand that growth comes from allowing people to make their own choices and learn from their mistakes. They don't try to shield them from this process by making choices for them as do many parents, for example.

Innovators in education believe that learning should take place in schools the same way: children should be allowed to choose their course of learning, pursue it in their own way and at their own pace, and adjust their course as needed. Education, then, is modeled after life itself, and it therefore teaches children how to move through life. In this model, there is no right or wrong way, simply many different choices. Mistakes aren't considered shameful but part of the process of learning. This teaches children to know themselves and to be responsible for themselves—to know what they want, go after it, and take responsibility for the consequences.

Innovators love to learn; they respect knowledge and intellectual competence, but they also realize that it is only one aspect of a human being and that the educational system should also nurture the emotional, physical, and spiritual aspects of ourselves. Their vision is the education of the *whole* child, education that not only teaches children how to think but also teaches them to honor themselves and all life.

Children must understand their emotions and the role they play in guiding them toward fulfillment. They need to know how to manage their negative emotions so that they don't sabotage their fulfillment or harm others. Emotional education is part of teaching people to function well in society and is as important as intellectual training. Those who are emotionally wounded or at the mercy of their emotions cannot contribute to society as well as they might, no matter how intelligent they may be; and many actually cause harm to society. If the schools do not take responsibility for this, who will? Not the family, which all too often perpetuates emotional abuse and pain.

The same can be said for nurturing the spiritual aspect of people. If children cannot see their lives as meaningful and purposeful, why should they strive to benefit society? When you realize that you are connected to something that is greater than yourself, you can work for higher causes in your immediate environment. Without this connection, you are apt to follow the selfish dictates of the ego, which is behind the struggle and suffering in your world. There can be no peace in the pursuit of selfish goals; there is only peace in working for good. Strong societies are built on cooperation, love, and a desire for mutual benefit. The weakest societies are those of service-to-self worlds where hatred, competition, and oppression abound. Therefore, if you are to have an educational system that supports society, it must teach children that life is meaningful. And it must teach the values of love, forgiveness, cooperation, and dedication to good.

Many high-minded Star People are here to teach these things to your children. Many are simple people whose reach does not extend beyond their local schools. Many, however, work in administrative positions in schools that are floundering under the stress of the current times. We wish to encourage these individuals to be brave enough to make a difference, to have the courage to implement their new ideas and to fight for what they know is needed now. You are the ones who can really make a difference in the future. You have the power to make change. That is what you have come here to do, so let nothing stop you! The positive forces must be as determined to fulfill their mission as the negatives. It is up to you.

Star People in Politics

Star People will play a significant role in the politics of the future. The United States is on the verge of much political change and will lead the way in political reform worldwide. The political system of the United States will be the most effective system in the world once certain reforms are made. Many of these reforms are being examined now, but they will not be implemented for some time. The wheels of political change turn slowly.

The world is not quite ready for the kind of political leaders that will lead it in the New Age. These individuals have not even been born yet. In the meantime, certain Star People are preparing the way by pointing out flaws in the system and indicating the need for change. That is the first step. After general acknowledgment that certain reforms are necessary, then others will lead these changes. For now the roles of Star People in politics are primarily those of critics of the system and harbingers of change. They are not people who have actual political power at this time.

The needed reforms relate to eliminating all that stands in the way of the system becoming more responsive to the individual. As it is, the system answers more to special interest groups and the moneyed elite than to the people. These reforms include abolishing the electoral college, lobbyists, and even representatives. With computers in every home, it will be possible in the future for people to educate themselves on the issues and to register their opinions without going through a representative. This will help eliminate the apathy that currently distances people from political involvement. People will have to take direct responsibility for their government rather than leave it to the politicians. That is how it was meant to be. The politics of the future will be a politics of the people.

One of the most important changes will be the decentralization of government. More power will be given to communities to make their own laws, which will allow minorities to create the type of environment and life-style they choose. People can then choose which community to join. This will allow for more diversity in life-style and more peace within your communities. When people feel they have more say in their community and it reflects them more closely, they will work to maintain and improve it.

As it stands, many communities have abdicated their responsibility to the police, schools, social workers, and other government agencies. It has never been the responsibility of these institutions to create community. People create community. The false belief that someone else will fix the community besides those living in it is what is behind the deterioration of your neighborhoods. People will begin to see that this is their responsibility. Fortunately, in the future, people will have the time they need to devote themselves to this task.

Your social structure is changing so therefore your politics must change. The gap is widening between the rich and the poor, and the wealthy are the ones in power. If your system functioned as it should, this would not be happening. In a truly democratic system, the people are in power; the rich and poor build a society together that serves both. Most of the poor in your country do not want to be taken care of; they want to be self-sufficient. This is difficult, however, with the disparity of wages. The working poor are being used to provide cheap services so that the wealthy can be served.

The low minimum wage is behind the disparity between the rich and poor. The greater the difference in wages, the greater the difference in life-style. If you want to do something about this disparity, you must change the wage structure. However, those in power do not want it to change.

This change would not mean a loss of wealth but a more equitable distribution of it. The benefits would include more harmony, less crime, and a transformation of society. Imagine what it would be like if so many were not caught in poverty. Imagine what it would mean for your society to have less fear, hatred, and oppression. Could so simple a change as raising the minimum wage and putting a cap on earnings really bring greater happiness and social progress? This is what we suggest.

The United States holds individual freedom above social progress and welfare. The problem has been that your ideas about social welfare are misguided. There are other ways to design your society that honor both the individual and the welfare of the whole. Your rich are beginning to feel the effects of this disparity in wealth with their lack of freedom to move about safely. Something must give. The situation may worsen before you are willing to implement a change in your wage structure. Star People will play a role in calling for and instigating these changes.

Politics must once again become an honorable profession. This means that those entering politics must be honorable. We are not implying that this is not the case now, but the system does not reward honesty and service to the people; it rewards special interests and tempts those in power to compromise their honor. These temptations need to be eliminated and your representatives held accountable. So not only must honorable people enter the system to change it, but people must take responsibility for their government once again. You have the power through your representatives to influence and instigate change. If you don't exercise that responsibility, someone else will—such as the special interest groups. Get involved! Express your opinion and work for political reform.

Star People in the Creative Arts

Many Star People are drawn to the creative arts because they provide an opportunity to bring through ideas, inspirations, and healing. Star People have ready access to altered states of consciousness that allow them to be channels for creative ideas, music, images, poetry, songs, and dance. The result usually presents a fresh perspective, uplifts, enlightens, or heals.

Occasionally their work represents the Shadow side of the self or the culture as a way of forcing people to confront and clear it. These creations are distinctly different from those produced by negative Star People and others of a negative orientation. The difference lies in how they make you feel. Positive creations may leave you shocked and open to change, but they also leave you hopeful and with a vision. Negative creations leave you feeling fearful and powerless.

It is unfortunate that many of the initial movies and books about extraterrestrials instill fear and dread. This is due not only to negative influences but also to your own predisposition toward life's violent and darker side. Both positive and negative visions of extraterrestrials are being created now, but those produced are mostly of the negative type, with the exception of the movie *ET*.

This will change as more Star People gain positions of power in the movie industry. For now, movie producers will continue to play to the public's fascination with violence and the occult rather than the Light. It is not considered chic to extol the Good, the Light. It smacks of religiosity, which is feared and hated by many due to their upbringing in fear-based religions. It is unfortunate that the search for goodness is confused with religion, since it is really a natural human drive.

Many artists and writers are receiving images of extraterrestrials. Many are unaware of the source of these images or how real many of them actually are. Many science fiction writers, for example, see themselves as simply having a very active imagination. Some of these images, though, are remembrances of actual encounters with extraterrestrials during sleep, and others are images fed to them through their unconscious by extraterrestrial beings. These images help prepare you for the time when extraterrestrials become a reality to you. Having already seen some of their faces will ease your shock at their strangeness.

This is not as innocent as it may seem, however. The negative extraterrestrials are trying to scare you with these images; they use artists, writers, and movie producers to further their own interests. They want you to be afraid of extraterrestrials because that gives them power over you. So, although introducing pictures of extraterrestrials can help ease the shock, if all you receive are scary pictures and stories about extraterrestrials, your reactions to all of us will be negative. Two forces are at work here: one uses artists and others to pave the way to more positive relations with extraterrestrials, and the other uses them to frighten you and pit you against extraterrestrials.

Many artists also receive images of other worlds and dimensions, either as a result of their own interdimensional travels at night or from visions they receive in their waking hours. Those who can receive these images are highly developed souls who use their psychic abilities to help humanity. Since you are becoming intergalactic and since you are all interdimensional anyway, even though most of you don't know it, these individuals are merely ahead of their time in presenting this information to you. They are harbingers of what everyone will experience in the not too distant future.

The increase in the depiction of angels and heavenly dimensions is a positive trend that uplifts, provides hope, and helps people transcend their everyday difficulties. Angels strike a deep, archetypal chord in people, and although they have religious overtones, they have been one way for positive forces to reach humanity. Angels are just one type of positive being that interact with humanity and can be considered extraterrestrial. However, extraterrestrials are not generally considered to be as positive as angels, although many of them operate from such a high level of service as to be angelic.

Many Star People act as channels for music from other dimensions. This music not only awakens remembrances of these dimensions and assists other Star People in waking up but provides sorely needed healing. This healing is accomplished primarily by raising the vibration of the listener and clearing blocks in his or her energy field. These blocks can re-form, but repeated raising of consciousness this way eventually leads to a more permanent clearing. Repeated experiences of a higher vibration make it easier to find your way back to this level of consciousness and to stay there for longer periods. In this way, music can lead you to higher states of consciousness and help establish these states in your daily life.

Not all music uplifts this way, of course. Some music perpetuates the ordinary states of consciousness produced by the lower emotions: jealousy, anger, fear, hatred, and self-pity. Negative Star People are sometimes involved in producing music that supports these lower states. The music and lyrics perpetuate fear, exploitation, hatred, and prejudice. Music, like other media, is used by both positive and negative extraterrestrials (through Star People and others) to further their respective goals.

Dance can serve a similar role as music in raising your vibration and promoting healing. Many find that, if they allow themselves to move freely to higher vibration music, their consciousness becomes heightened, their energy blocks are cleared, and they receive insights and inspiration. Many Star People have discovered this and are using dance to enhance their own lives. Many also are teaching dance to others as a way of waking them up.

Some people actually channel their dance and other movements. Extraterrestrials and other beings can animate the body of someone who is a medium and use it to express themselves. Here again, the expression may be positive or negative. These are just some of the many ways nonphysical beings communicate with people on Earth.

Star People in the Media

Star People have not infiltrated the media yet because the media reflect consciousness rather than transform it. That is not to say it will always be this way, but rather that they're not being used to inform and transform as they might. The media cater to your baser instincts; they feed your lower self while starving your spiritual self. When people of a higher consciousness gain positions of power within the media, this will change. Then the media will become a powerful force for good.

Right now, Star People are on the periphery of the media, dreaming their dreams of influence but finding few inroads. Transformation of the media, like most changes, will happen slowly and gradually, with certain individuals paving the way for greater numbers in the future. Those of a higher consciousness are in the minority and for that reason must take a back seat to mass consciousness. However, Star People have the advantage of being more intelligent, creative, and gifted than the ordinary person; therefore, they are bound to stand out among the crowd and be heard.

Most Star People are longtime fans of science fiction or anything having to do with outer space, so it is no surprise that this is where they have their greatest impact. They are also the ones interested in the paranormal, UFOs, out-of body travel, and ghosts. They are interested in these things because most have experienced them. Star People not only try to understand these experiences but seek others who have had them. Paranormal experiences are commonplace to them yet their culture denies their reality. They KNOW these things are real, though. They aren't trying to convince others so much as trying to come to grips with a culture that denounces them.

America's puritan ethic views paranormal occurrences with skepticism and downright criticism. The way the media deal with the paranormal is to represent it to the public through the consensus' distorted lens of skepticism and fear. Nevertheless, the mere presence of paranormal subjects in the media paves the way for a more positive depiction of them in the future. Before long, the media will disengage from the scientific viewpoint and begin to reflect what many people believe—that the paranormal exists. As more people have paranormal experiences, the evidence for them will become irrefutable.

Paranormal experiences are becoming more common; it's not just that people talk about them more freely. They are more common because there are more advanced souls who are capable of experiencing these things than ever before, and they trigger these experiences in others. Extraterrestrials also are more involved with people and your planet than ever before, and this affects people's openness to the paranormal. The more open to the paranormal that people become, the more experiences they will have. Paranormal experiences are readily available. Everyone has them while they sleep; they just aren't remembered. Even dreams are paranormal experiences. You are multidimensional beings and the more you open to these truths, the more you will experience them.

Star People in Science and Medicine

Science and medicine are relatively conservative fields, and Star People in these fields are there to question established tradition. Some question how things are done while others look for new solutions. Most often, you will find Star People in research, seeking cures and developing technology to further humanity.

Star People are humanitarians, but they are also adventurers and pioneers. They are not afraid to challenge the establishment if it is unwilling to back their dreams and goals. Most will play the game only up to a point. They will not compromise their integrity. This is why many don't get the education they need or why they may not achieve their goals. Their independence and unwillingness to compromise often work against them in a world that demands compliance before it bestows power.

Star People may carry around a sense of superiority, which can sabotage them. The ego often translates knowledge of who they are, or the sense of being different, into a feeling of being special and not having to follow the rules. Their intelligence and talent usually give them an edge, but unless they also are willing to play by the rules, their gifts may get them only so far. Too often they discover that they should have been more disciplined and more compliant. In the end, they have only themselves to blame for being unable to influence the world as they would like. This arrogance is a potential pitfall.

Medicine and science have little sympathy for those who are unwilling to follow established rules. You prove you are one of them by abiding by their rules and procedures. Those who don't are considered outsiders and their ideas dismissed. Therefore, to make an impact in these fields, it is usually necessary to become a respected insider. This is always a challenge for Star People. Fortunately, besides being rebellious and innovative, most Star People are also determined to accomplish their goals and many do succeed.

The future will see far more Star People in science and medicine. The more open these fields become to new ideas, the more they will be flooded by Star People. Meanwhile, a few pioneering Star People are breaking ground for others to follow. These groundbreakers are especially resilient souls who have chosen this challenge knowing the difficulty but feeling confident that they are up to it. Many have performed this same service on other planets and have become expert groundbreakers through experience. They travel the universe acting as catalysts for change, a role they find most rewarding. They leave what comes next to others who are more skilled at implementing change once it has been initiated.

Besides being courageous, these groundbreakers are intelligent. Because these fields require a high level of intelligence, these Star People must be smart. Most are considered geniuses by your standards, but this level of intelligence is really quite ordinary for a Star Person. It does, however, require a special aptitude and training in the sciences.

In serving Earth now, some of these Star People are balancing karma they created here or elsewhere by misusing their intellect or scientific knowledge. We should mention that there are also negative Star People in the sciences who put their intelligence to use without regard for the consequences to humanity. They are responding to a drive to serve themselves and may also be being influenced by negative extraterrestrials who use them to promote their goals.

Science has always attracted negatively-oriented individuals. One of the characteristics of the advanced service-to-self is a high intelligence but without the spiritual and emotional development to balance it. These individuals have caused damage to civilizations throughout the universe. Eventually they will make amends for this damage once they choose the path of service. Thus, some of the positive Star People working in the sciences today were once negative and are now repaying their debt.

Star People as Healers, Psychics, and Telepaths

There are more Star People acting as healers, psychics, and telepaths than in any other profession. This is because the need for healing is so great and because psychics and telepaths provide extraterrestrials with direct access to people on Earth. Except for rare direct encounters, psychics and telepaths are our only means of communication with people on Earth. We use them more often than encounters to communicate with you because this method is less shocking and reaches more people. Besides, most of the direct encounters are either repressed or not believed.

Dreams are the next most common way that extraterrestrials communicate with you. This happens nightly so, whether you are aware of it or not, you are undoubtedly receiving communication of some sort from extraterrestrials. The type and source of communication depend on your interests, your life task, your level of consciousness, and your orientation.

Star People often receive training in their dream state to help them carry out their life task. For example, many healers receive training in healing techniques while they sleep. Some are already master healers, but all healers need help in regaining their innate knowledge and putting it into form in third density. It is one thing to have used these techniques in fourth density and another to apply them in third density.

Some healers receive training directly through telepathy or other psychic means. Some even act as mediums for disembodied healers who work through them. Many of them practice their healing arts in other dimensions while their bodies lay sleeping.

Many of the advances in healing are coming through telepaths and psychics. They not only bring through information about healing and new techniques but provide diagnostic and treatment information, just as Edgar Cayce did decades ago to the marvel of many.

Some Star People who are mediums are channeling mudras (special hand positions) and other body positions that assist in energy clearing for themselves and others. Many of the body therapies and means for clearing the body are introduced by Star People who have a special affinity to the body and who share with others the knowledge that is given to them through their own bodies.

Star People are also involved in toning and other uses of sound in healing. This technology is well-developed on many other third- and fourth-density planets. Those introducing it on Earth use sound technology not only in healing but building construction and even travel. This field will grow enormously in the next 25 years.

Many cultures throughout the universe have discovered the benefits of using color to balance the aura and therefore promote healing. This, too, is a field that will flourish in the next few decades as more and more Star People unravel the mysteries of the human energy field, or aura. Understanding the aura is the next step in unlocking the secrets of health, and many Star People are already involved in this great discovery process.

As in other fields, Star People who are already in the health fields are questioning traditional practices and inventing new ones. Many are working to increase your understanding of death and the afterlife and to bring spirituality into your hospitals, especially to the bedsides of dying patients. The crisis of death provides an opportunity for both the dying and those left behind to experience their connection to Spirit. Star People help people die more consciously and help their loved ones accept death as a part of life.

One approach to this is publishing accounts of near-death experiences. A near-death experience is almost always based on a prelife agreement, thus the opportunity to return to life is rarely passed by. Many Star People and others have volunteered to have these experiences to awaken others to the spiritual realm and, in the process, awaken their own natural abilities. A near-death experience often endows the person with remarkable healing ability and psychic powers.

Star People are more evolved than the ordinary human and an awakened Star Person is a natural healer. Because they are like the ordinary human's future self, they have abilities that all of humanity will have and take for granted in the future. Healing by touch is one of these abilities, which Jesus Christ and other Masters demonstrated. Someday, everyone will have this ability. In the meantime, the ability to heal will awaken in more and more Star People.

Telepathy and other psychic abilities are also natural to Star People. However, not all Star People are telepathic or psychic, though most can access information from other realms more easily than the ordinary person. For some, receiving psychic information would interfere with their chosen life task. Some need to appear more mainstream and may choose to keep their psychic abilities closed down, except on occasion or when helpful to their work. In these cases, you can be sure they are receiving guidance and information in both the dream state and intuitively to help them with their task.

Star People often choose to limit their psychic abilities so as not to appear too different. If they allowed all their abilities to be expressed, they would have a difficult time fitting into society. They might even end up in mental institutions, which some do anyway. Your world is not very accepting of those who can perform miracles. But what you call a miracle today will be commonplace in the future. It is difficult to convince you of that, though!

Until science begins to explain these miracles, those with special gifts will continue to be misunderstood and even persecuted; therefore, Star People are wise to keep a low profile. They need to understand that change happens slowly and take steps to ensure that their psychic unfolding does not disrupt their mission. Of course, there is a place for the demonstration of miracles as proof that science doesn't yet have all the answers, and some Star People are here to point this out.

Chapter 4

Your Changing World

Change is accelerating for a number of reasons, including computers and other technology as well as the change in vibration. The planet's rise in vibration is causing a similar rise in vibration of people's energy fields, or auras. As the vibration of the aura is raised, blocks in the aura are released. In this way, the planet's heightened vibration catalyzes our growth. It stirs people's issues up in the process of healing them.

Blocks are energy structures that are stuck in the human energy field, or aura. They stem from events earlier in this or other lifetimes and hold the energy associated with some type of trauma. For example, a rape victim may fear the dark for the rest of her life, or someone who died of starvation in another lifetime may feel he can never get enough food. Blocks may also reflect negative beliefs held over time, which continue to reinforce themselves. For example, someone who believes that there is never enough, will see lack and not abundance, and his experience will be in line with that, which to him justifies the belief.

Prior to release, these blocks may be amplified and cause problems or crises in a person's life. As part of the process of healing these psychic wounds, many things can happen. People may consult psychotherapists and other healers to help them with emotional problems related to these wounds. Or, if these emotional problems manifest physically, they may seek help from physicians or other health care practitioners. More negatively, some may lash out irrationally, acting out their pain as their emotional wound is revealed. They may precipitate crises in their lives through violent or hurtful acts against themselves or others. And some may collapse for a while and let others take care of them in hospitals or institutions.

After the wound has been faced and steps have been made toward healing it, people often rearrange their lives to accommodate their new attitudes and sense of self. Thus, change happens first in the form of a healing crisis and then in some form of rebirth. As more and more people experience such dramatic growth, they activate and facilitate these types of changes in others, especially in their loved ones.

As more tools become increasingly available for growth and healing, and as turning to these tools becomes increasingly acceptable, more people are taking advantage of them. The pain in the world is great; the pain in individual lives is great. More than ever, people are aware of their pain and want to do something about it. The climate is finally conducive to healing, growth, and change; so change is happening!

Those who resist growth and change in these times will suffer more than they need to. Fortunately, there are many excellent models of people who have successfully overcome crises. Christopher Reeve is a good example. Many of those in crisis, such as many AIDS patients, are Star People or highly developed souls who chose this challenge not so much for their own growth but to model positive coping skills for others. Of course, serving in this way also accelerates their own evolution. But those around them also learn and grow.

Some people thrive on change while others resist it. This difference is due in part to beliefs and in part to differences in personality and soul age. Your beliefs will be essential in helping you through these times. Great strides can be made in spiritual development now. By spiritual development, we mean awareness of oneself as a spiritual being, which includes being aware of all as Spirit and having a relationship with Spirit. That is what is missing in your world today, particularly in Western culture. What a different world this will be when people are more aware of their spiritual selves!

Crises have a way of putting people in touch with their spiritual selves like nothing else. Crises are not necessary, of course, to develop spiritually; but many unconsciously create them to catalyze themselves. Older souls are less likely to need a crisis to stimulate this kind of growth, but many Star People nevertheless unconsciously choose this route as a way to wake up.

The whole world is waking up. It has to. You have reached a point where you cannot continue destroying your planet. Your unconsciousness is ruining your chances for survival; therefore, you must become conscious. This means taking responsibility for your actions, being aware of how they affect Earth and those around you, and making a conscious choice about whether to keep acting that way.

Until now, it has been easy to be unconscious because Earth's resources have been abundant enough that you could ignore the consequences of your actions. But now you are faced with polluted water and air, lifeless food, diminishing rain forests, the growing hole in the ozone layer ... we could go on. You are on the brink of planetary ecological disaster, yet some of you still do not see that this must stop.

The realization of this crisis will make it necessary for you to change your life-style and many of your social structures. For example, when you don't have enough clean water, you can't wash your clothes, water your lawns, or fill your swimming pools. That is one level of crisis—certainly an inconvenience for you—but nothing compared to dying of thirst or disease from unhygienic conditions, which is already happening in many places around the world.

Disease and death from a lack of clean water are not something Americans have seen or even expect to see in their lifetimes. But when people all over the world are already suffering this crisis, why are you so confident? How much longer do you think it will be before you are affected? There are certain things that your technology cannot create or even fix, and the lack of water is one of them. For you to face this crisis in the future, you will need help. Fortunately, you will receive some help from us in solving your problems but that is no reason to do nothing about them now. The lack of clean water is only one problem your world will face if you don't make some changes now. There are many others.

Our intent is not to frighten you, but to help you open your eyes and see that change is inevitable. The changes will be easier to make if you take action now. At this point, it is not even a question of whether you will face gradual change or less gradual change, but rather whether you will face rapid change or drastic change. That is your choice. Will you voluntarily initiate change now as fast as you can, or will you be forced to change by crisis? Regaining order and instigating change is much harder once chaos sets in. It is better to institute changes while you still have institutions.

Again, we are not trying to frighten you, but some of you aren't afraid enough. You should be afraid of losing your water. You should be afraid of losing your air. You should be afraid of losing your soil. Fear helps you survive but your technology has lulled you into a false sense of security about the environment. In your arrogance, you think any problem can be solved. Perhaps it can, but can it be solved in time?

Famine, Pestilence, and Disease

More people are dying from famine, pestilence, and disease than ever before. You may say, "but there are more people than ever before, so of course this is true." And you would be right. Is it right, however, that this happens in a world that can prevent these deaths? Why is this all right? Why aren't those of you who can do something about it doing something?

To us, these questions are obvious. We would not allow such suffering on our worlds, because we take to heart the concept of being "our brother's keeper." You, as yet, do not see yourselves as being responsible for ALL human beings, only those of your family or your community, or possibly your nation. You are not yet thinking globally. That is the big jump you will make soon, because you will have to.

Because you do not think globally, people die and valuable human resources are lost. What if Einstein had been born in Africa instead of the West? Is it so inconceivable that a genius, perhaps even a savior, could come from a third-world country? It has happened in the past. Is an American, an English, a German, or a Japanese life more valuable than a Somalian life? Are you willing to make that judgment? You make it every day by not taking care of the world's poor and suffering. You choose your swimming pools and BMWs over lives. Their suffering is real; their deaths are real. This is not a movie that has a happy ending. It is happening in your world yet you pretend it's not happening. This would be incomprehensible to us if we hadn't done the same thing earlier in our evolution. That this is happening on your planet without your doing anything about it reflects the level of your evolution.

We are going to help you get over this! We have lots to teach you. We will show you how valuable each part is to the whole, what the world could be like if you shared your planet's resources. We can show you this in movie form, a format that so appeals to you. We can show you pictures of probable future realities, both positive and negative. We will show you what your world would be like, given certain choices, to prove to you that your future is up to you. You determine it with your choices. Right now your choices are leading you deeper into destruction, disease, pain, and suffering.

You can change this by changing some of your beliefs, beginning with the belief that you don't need to take care of those who suffer in your world. You're right; you don't need to. It is your choice. That choice, however, can lead to your own downfall because not only does it slow your spiritual growth, more importantly it perpetuates a world full of suffering. You perpetuate suffering by not attending to it. The only way the world can be at peace (which is our goal for your world, and should be yours) is if its resources are shared equitably and if you begin to live like one world rather than separate nations—a tall order, we know—but this is the kind of change that must and will come about on your world. This is the kind of change we are here to facilitate.

The famine, pestilence, and disease in the world today are symptoms of your own spiritual disease. These problems demand action, but this means opening your heart. Doing that would require a leap in your spiritual growth! Well, the planet is ready for that. It needs that. Famine, pestilence, and disease are the manifestation of your own selfishness, hatred, and prejudice, which you must overcome before your world is free of such suffering. It is hard to look at these things and it will not get any easier until a shift happens within you. Until that shift, you will not be able to face those who suffer. After the shift, you will recognize yourself in them and reach out your hand in compassion.

Climate Changes

The world's climate is changing. There are a number of reasons for this, but many are man-made. The environmental impact you have on the planet aggravates many natural changes that simply reflect the planet's own evolution to higher consciousness.

It is not unusual for a planet's climate to change. What is unusual is for it to change drastically and suddenly. Usually, a certain homeostasis exists in which minor adjustments maintain an overall balance. Most climatic change is gradual; sudden change such as ice ages, annihilates whole populations of living creatures. Human beings can withstand a relatively narrow range of temperature. Raising or lowering the average temperature even ten or twenty degrees can have an enormous impact on the human population, not to mention other creatures.

It is clear that Earth's climate is changing yet it is very difficult to predict, even for us. We are studying your climate as we seek a better understanding of your planet so that we can help you. It is clear that some places will become too hot to inhabit and others will become much colder. Where these places will be is uncertain, but you will see more extremes in climate and weather: more hurricanes, tornadoes, floods, and droughts. Where they will occur, again, is not clear but they will be worldwide.

There is no doubt that hurricanes and the like will increase—they are on the increase even now. What their effects will be remains to be seen. The wheels have already been set in motion and you can only do so much about the weather now. What you can do is be prepared and learn to share your resources with those who are hit by tragedy. You could be next. You will discover the wisdom and blessing in "doing unto others as you would have them do unto you." Crises can bring not only a nation but a world together. Crises open people's hearts. They develop compassion. They bring people together.

Geological Changes

Planet Earth is experiencing growing pains that result in increased volcanic and earthquake activity. These, too, are on the increase. Much has been said already by others about these changes and why they are occurring. Most of all, we want to point out that they serve a purpose not only for the planet but also for you in having to cope with them. Like the weather, they provide challenges that can help bring your world together. In case you haven't realized by now, that is on Gaia's agenda now. Humanity is to unite as one world, and Gaia is doing her part in bringing this about.

Gaia is going through some necessary physical adjustments to accommodate a shift in consciousness. These adjustments will also shift your consciousness. It is a wise plan, although rarely has a world gone through such a rapid transformation. That is one reason we are here. You need the extra help. We are helping Earth make the necessary adjustments. We have the technology that will reduce the disruption to the human population. We are like midwives to your Earth. We have served many other worlds in this capacity, although every world is different. So, we are not here just for you but for the planet.

Pole Shift

There will be a pole shift as part of the planetary changes. It will affect the climate throughout your globe, but to what degree remains to be seen. The extent of the pole shift and its effects are uncertain. It appears that the effects will not be as dramatic as once thought. That is the good news. The bad news is that it will affect life on your planet as whole populations will need to relocate to accommodate the climate changes.

When this will occur is also uncertain. We think that it is unlikely to occur before 2012 but it is too soon to say, and that date is just the best current guess. When the time approaches, we will know.

Coping With Change

We will help you cope with the changes, in accordance with what you ask for and with what is in your highest good to receive. We will not transgress your free will by interfering where we are not welcome. We will not give you everything you ask for, however, because that might not be what is best for your evolution. We will establish a partnership with you, but like all partnerships, there must be an equal exchange.

What we will ask of you in return is that you set down your nuclear weapons and all other arms and begin to live in peace. So, you see, what we want in return is not what the service-to-self extraterrestrials want. We ask for nothing in exchange, except what is best for you. We are here to serve you and your highest good. That is not true of the negatives, who will demand concessions in exchange for their so-called help. They do not want to help you, so what comes in the guise of help will not be help. What the negatives want from you is your allegiance, your obedience, your power. It will be important for you to pay attention to the motives of those from whom you seek help.

Now, some of you may say that to put down your arms places you in jeopardy, makes you vulnerable to us or to the negative extraterrestrials. The truth is that if we had wanted to conquer you by warfare, we could have done so long ago and could still do it whether you have nuclear arms or not. The negatives didn't take that approach either because they knew they have us to contend with if they did. Besides, they prefer to conquer through fear rather than with weapons. That doesn't mean, however, that they won't try to scare you into submission by threatening you with their weapons.

Cooperation is the way to get through these changes, not just cooperation with us but more importantly with each other. You will learn the power of joining together with others to overcome adversity. Then you will see each other differently. You will see each other not as competitors but as co-creators, and together you will rebuild your societies. The only other option is anarchy and degeneration into service-to-self. In overcoming your differences and working cooperatively with each other, you will build a new world. So you are faced with only two options: competition and anarchy, or cooperation and peace.

Chapter 5

The Role of Evil in the World

In your world and in many others, good evolves from evil. By evil, we mean actions that harm others. By good, we mean actions that evoke love. When evil is done, the perpetrator experiences the result, either immediately or in another lifetime. No evil act goes unbalanced. The Law of Karma always creates a circumstance by which the lesson of love is learned. You may be wondering why this is. Why have this as a planetary premise? Why not just have good evolving into more good? Well, that already exists on other planets. That is what higher evolution is about. But first, many beings choose to experience evil as a way of evolving into love. It is simply one way the Creator seeks understanding.

The Creator explores life through creating circumstances by which to evolve. You, as offshoots, or expressions, of the Creator are the Creator's way of exploring the evolution of good from evil. The Creator is enriched by this exploration, by this experiment.

Not all the worlds that evolve this way are like yours. Not all of them, for instance, have beings on them who have feelings like you do. Your species is unique to all creation. Although there are similar creations elsewhere, no one in the universe is exactly like human beings on Earth. The Creator is exploring what it is like to have your unique combination of free will, feelings, sentience, physical form, and spiritual development. It is a difficult combination, to be sure, but one the Creator values for its potential for growth and understanding.

Wherever evil exists, the opposite also exists. Even though there are worlds that are based solely on service-to-self, goodness and love still exist and evolve on those planets. Those who develop love no longer choose to incarnate on those planets. Learning to love makes it possible for them to graduate to planets where they can continue to love and be loved. Love is able to develop on service-to-self planets because compassion develops from pain, and compassion leads to love. Therefore, those who are victimized on those planets eventually develop compassion and in turn love. Those who victimize on these planets reincarnate as victims and thus learn compassion. They circle round and round as victim and oppressor as long as they want, but eventually they choose to feel compassion, which releases them from their prison of oppression and pain.

The victims on these planets are often stuck in victimhood, anger, revenge, hatred, blame, martyrdom, and self-pity for several lifetimes or even longer before their compassion is sufficiently developed to make a different choice. Thus, both oppressors and victims on these planets have their lessons. Both roles are played out until the person is ready to move forward toward love.

This same dance of victim and oppressor is happening on Earth today and has happened throughout its history. Earth has always been a place where the duality of good and evil is played out. In part, it is being played out here because your creators, the extraterrestrials responsible for your creation, were enacting these same roles on their home planets where this drama was going on. They created you in their image, not because they wanted you to experience the same pain they have had in their evolution, but because they could not avoid recreating their struggles within you. They, too, had the same combination of free will, emotions, sentience, and physical makeup that make for this set of challenges. Because of their makeup, they could not help creating circumstances on Earth similar to those they have created everywhere they have settled.

Some of them hoped to create a better civilization on Earth, one free of the conflicts and oppression of other planets. Some believed that with proper guidance from them, you would not have to experience the same difficulties they did in their evolution, that somehow you could benefit from what they learned and skip over some of the pain, disease, and tragedy that accompanied their growth.

On the other hand, some of your negative extraterrestrial ancestors had every intention of continuing their victim/oppressor drama on Earth. They were working against those who hoped for better for you. The negatives worked against the positives by genetically engineering you for a continuation of that drama, and by reincarnating on Earth and playing out that drama themselves. Many from service-to-self races have reincarnated on Earth and continued their drama here. This goes on today as it has throughout your history. There is an especially strong contingent of service-to-self on your planet today. The forces for both good and evil, or service and service-to-self, are gathered in great numbers to play out the finale to this drama.

In the very near future, Earth will no longer serve as a place for service-to-self to reincarnate; therefore, it is inevitable that those in service to the planet will win out in this drama. But that doesn't mean that things won't be intense in the meantime. The battle for the soul of Earth, so to speak, is on, and you are here to take part in this important time. Whether you sit back and watch or actively contribute is up to you. But many of you know you came here to oppose evil and serve the good, and you will not be able to simply stand by and observe as evil proliferates.

What does it mean to oppose evil? This is something that those on service-to-self planets are learning. Once they learn the way to oppose evil, they no longer are caught in the victim/oppressor drama on these planets. They learn that you do not oppose evil by joining evil in violent and harmful acts. You do not fight evil with evil, but with good. Good is a powerful weapon, far more powerful than evil. Evil only perpetuates suffering, while good heals it. Those who perpetrate evil suffer themselves.

They also suffer from a lack of spiritual understanding. So, another way to fight evil is through instilling understanding. Those who perpetrate evil do not understand certain basic laws of life. Even if they have heard of these laws, they refuse to accept them; they refuse to believe. They suffer from a lack of faith, a lack of vision, and a lack of hope. They hold mistaken beliefs that keep them trapped in the victim/oppressor drama. Some of them are stubbornly attached to this drama and their pain. They are either caught up in self-righteousness and revenge or they have given over all their power and are lost in hopelessness and helplessness, believing there is no way out.

One law of life those caught in drama refuse to accept is the Law of Karma. Those who are the oppressors do not believe that the harm they do will come back to them and, in fact, their societies do not teach them that. Their societies teach the opposite: that crime pays. Those who are victims not only believe that they should be victimized but that they deserve to be enslaved because they lost the battle against their oppressors. Their only hope, therefore, is to beat their oppressors—to become the oppressor instead of the victim. And so the drama goes on.

Does this sound familiar? It is what your wars are made of: oppression creating pain and retaliation, creating more pain and oppression. The way out of this is to stop the oppression, to stop the retaliation. But who will stop if anger, revenge, self-righteousness, hatred, martyrdom, and self-pity are celebrated. As long as these emotions are allowed to reign, peace and love cannot prevail. The way out of this vicious circle is spiritual understanding. This, alone, can break the cycle. So, there must be an understanding of the Law of Karma. This law teaches love, not hate—that hate produces more hate, and love produces more love.

Something else that is lacking in service-to-self worlds is a respect for love. Love is misunderstood as weakness on these planets. Loving someone makes you vulnerable to them, and that is potentially dangerous on a service-to-self planet. The relationships that form on these planets are ones of camaraderie, either in crime, rebellion, or revenge. Any partnerships that develop are centered around such a cause, or based on shared values and beliefs. This is not unlike relationships between younger souls where the capacity to love and share intimacy has not yet developed. Intimacy requires trust, which is a liability in a service-to-self world. Partners on these worlds trust only enough to work together, but it is not the level of trust that exists in a true love relationship where there is a giving of oneself to another. People do not give themselves to another on these worlds. Self-reliance and self-protection are paramount, even within relationships. So, on service-to-self worlds, relationships are formed for mutual protection and support, but they do not serve as vehicles for the development of love.

In a world where there is little love and what love does exist is looked upon as maudlin, sentimental, frivolous, and weak, children grow up wounded and warped. In these worlds, generation after generation is raised to hate and hurt. The cycle continues because no other way of life is known. There are no models on these worlds for other ways of being. You are most fortunate to have both the positive and negative operating in your world. So, although you see this same cycle of hate on Earth, you also know it doesn't have to be this way.

Knowing the positive, it is easy to see the folly of the negative. That is why Earth is, and has been, such an important training ground for those coming from service-to-self planets. They are reformed here, but only after agreeing to give the arrangement on Earth a try. Many stay on service-to-self worlds for a very long time before agreeing to try something new. Many are afraid they will be weakened by being exposed to a loving environment. They don't know what a loving environment is, so they really can't evaluate it. Others are just more comfortable with the known and are unwilling to try something that is unknown. Therefore, some stay stuck on these service-to-self worlds for eons.

Your world has served as a place to teach these individuals about love. In the meantime, they teach others about love through the contrast they provide. They demonstrate for those in service on Earth where service-to-self leads, and those in service show service-to-self a new way of living. Those in service on Earth don't mind serving in this way. On a soul level, they have agreed to come to a world that has both positive and negative to evolve and to help others evolve.

There have been times in Earth's history when service-to-self has prevailed and other times when love has prevailed. The balance has shifted back and forth as one extreme leads to an awakening of the other. Today, the polarization is intensifying, which intensifies the teaching. Those in service-to-self are succeeding in creating havoc in this world, while those in service are summoning their forces toward reform. The negative forces first seem to succeed, then the positive forces come forward to resist them. When the positive forces are asleep in their complacency, the negative forces rally once more. This cycle has occurred repeatedly throughout history. You are at a point where the negative forces have reached their apex. They will not gain much more control before the forces for good make a final stand. A confrontation is at hand.

The extraterrestrials involved with you follow the lead of those on Earth. The negative extraterrestrials assist and encourage negativity on your planet, while the positive extraterrestrials wait for you to engage your will against the negatives before they offer their help. You see, you must show a willingness to oppose the negative, a willingness to have our help before we will intervene. You must first choose the Light, and then we can help you. We are waiting for you to choose the Light—not only for you to choose it but to take steps toward opposing the Dark. Universal Law requires that you must prove to us that you want good to prevail; that is how you can summon our help.

You are entering a period of great change and during such times anything becomes possible. The doors of change open up new possibilities that could not even be conceived of in more stable times. The negative forces know that change opens the door for them too. They are waiting to take advantage of the instability of the next several decades. You will have to be very discriminating in the choices you make during the times ahead. If your choices are shortsighted and based on fear, you are bound to lose your way. You will need men and women of vision, courage, and integrity.

You know what the Light looks like and what unconditional love feels like, so you have no excuse not to choose it. You will be continually challenged to choose the Light over the Dark in the next few decades. The more you choose the Light, the sooner your world will become love-based and no longer a stronghold for service-to-self. It will not be long before those in service-to-self will be directed somewhere else to learn their lessons and your world will have graduated to a fourth-density positive outpost.

Chapter 6

Our Plan and How We Operate

Just as you have plans by which you run your lives, so do we. These plans, like yours, are changeable and need to be revised from time to time. Still, a plan is useful in achieving one's goals. We of the Confederation of Planets have a plan of action for our involvement with you. Other extraterrestrials have their plans also. Each individual within the Confederation has certain assignments in keeping with this plan. Their fulfillment of these assignments affects the overall effectiveness of the plan. Sometimes they contribute information or insight that makes it necessary to adjust the plan. Just as your plans take different turns according to the free will of those involved, so do ours.

We cannot control your choices, or the choices of other extraterrestrials involved with you. We can only try to influence you. You are the ones with the real power because you have the final say in what you will do. Some of you have more power in the world than others, although those of you with power are not necessarily the ones capable of making the best decisions. Conversely, some of you without power are wise, but have little means for influence. Our strategy is to work with both those who have power and those who do not. We work with those who have power for obvious reasons. Why we work with those who have little power may be less obvious, so we will explain.

Many Star People have special gifts and awareness but little power. They have little power usually because they are so different and sometimes because they are too rebellious to seek power or they just are not interested in it. Star People are not interested in power in the usual sense, although they are interested in spiritual power. We, too, are interested in their spiritual power and make use of it.

Many Star People have power in third density because of their special gifts and understanding. They have a powerful influence over people's lives. This kind of influence is important to us because the revolution of consciousness takes place within individuals on a grass roots level. This revolution of consciousness is a movement that is building and will be responsible for much change in the future. We seek to shape this movement by influencing those leading the way. These individuals may never be leaders in the world, but they will influence the kind of leaders you will choose in the future. You see, your politics reflect the majority consciousness. When this changes, your leaders will change, so influencing consciousness is a way of effecting political change.

We also work with world leaders and others in important positions throughout the world in the same way we work with others: through their unconscious, through their dreams, and through others in their lives. We can create circumstances and incidents that bring about certain realizations. We seek to change their way of thinking by arranging events in their lives in ways that teach them and bring them into alignment with a certain perspective.

You may be interested to know more about how we do this. You may think that events are random, but there are no accidents. Events happen either as a result of free will or because they are part of the soul's plan. We and other higher level beings are instrumental in carrying out the soul's plans of those in third density. But even when events are orchestrated, do not think you are puppets; you choose your responses to events. Nevertheless, most events are simply natural results of your choices.

We are not the only ones orchestrating events from these levels; negative extraterrestrials do so as well. While we influence you in ways that will help you fulfill your soul's plan, the negatives work to serve themselves. Let's look at how they operate.

Most events are determined by free will. The free will of most human beings is just that—very free; it is easily influenced. Most people do not have their will and awareness developed to a point where they are not easily swayed or manipulated. It takes someone with a strong awareness and sense of self to hold to a certain course. Most people are not that clear about what they want. They often take the path of least resistance, which usually means following the dictates of their unconscious or other people's suggestions. They are not only influenced by old patterns stored in the unconscious (which cause them to repeat what they've done in the past), but by suggestions implanted in the unconscious. Hypnosis is one way of programming the unconscious, as you have discovered. But what you may not realize is that nonphysical beings are actively programming people's unconscious minds.

Some of these less evolved nonphysical beings are programming or influencing the human unconscious in ways that are not at all helpful. For example, discarnates who have not found their way to the Light may stay attached to a living human being for one reason or another and influence him or her by their presence. Some of these beings are well-intentioned and some are not. Some actively try to cause trouble and some do not. A discarnate who is addicted to alcohol, for instance, is likely to be trying to influence someone to drink. On the other hand, discarnates who die of illness and are lonely for a loved one may be just trying to stay near, without realizing that they are imposing their own previous health problems on the loved one through proximity.

Another level of influence are those devoted to service-to-self, whose purpose is to create pain, fear, self-doubt, conflict, and illness. They study someone's weaknesses and move in when they perceive an opening. The way to protect yourself from their influence is to work on eliminating your weaknesses and to maintain high-minded thoughts and positive feelings. No one is able to do this consistently, however, so even the most evolved of you can fall prey to these negative forces.

Negative forces can aggravate any negativity you feel and make it difficult for you to shake it off. Once they have moved in, it takes diligence to keep from wallowing in negativity, whether it be jealousy, hatred, guilt, remorse, self-pity, anger, or sadness. When you find yourself trapped in these states, realize that it may be due to outside influences. Then call on help from positive forces. Remember, you must ask for help, for unlike the negative forces, we do not infringe on free will by interfering where we are not wanted.

You see, many of you cherish your negative states. They may be comforting in their familiarity, or they may give you just the excuse you seek to do nothing, to avoid taking responsibility for your life.

CHOOSE CONSCIOUS MIND OVER UNCONSCIOUS EMOTIONS,
ET CONTACT *FEELINGS, FORCES, ENERGIES — — NEGATIVE ENERGIES*

We need to know you are willing to feel good again before we will help. When you do ask for our assistance, most of you will be able to feel it immediately, both energetically and in a shift in your mood. You may still have to work at not sliding back into those states, but we can also help with that.

Some of you who have persistent negative thoughts might be plagued by a negative entity who has attached itself more permanently. In these cases a clearing is necessary. This is done by many healers these days and is more simple than you might imagine. Whatever it was within you that allowed this attachment must be healed, however, or another entity may replace the old one. That is why it is useful to have a healer help you with this. Alternatively, you can clear them yourself simply by speaking to them with great intent and telling them they are no longer welcome and must move on into the Light where they belong.

Much of the negativity in the world can be traced to this type of negative influence. Christians refer to it as Satan and his legions. Yes, there are devilish forces out there trying to undermine you and all that is good. There is a battle for your soul going on! But this is nothing new. It has always been this way on Earth. Your challenge is to recognize that this is happening and to use your will more consciously. Most people are tossed to and fro by one influence and then another because they are unaware of their thoughts and feelings. But this doesn't have to be. You are reaching a state in your evolution where you are becoming conscious and need not be at the mercy of unconscious forces—but you must choose consciousness!

Positive forces also work with your unconscious. You are free to choose whether to be influenced by these forces or by the negative. The negative voice from the unconscious is louder at times, such as when you are filled with self-generated negativity. Feeling anger, for instance, sets up a resonance between that emotion and any negative energy in the unconscious, which amplifies the negativity. The negative voice becomes a choir of negativity, drowning out the positive voice of peace, love, reason, hope, and faith. On the other hand, when you resonate to the positive, the voice heard from your unconscious can become like a choir of angels, inspiring you to love, joy, and exultation. Both voices vie to be heard; it is up to you to choose which you will listen to and then turn up the volume.

ANGELS VS DEMONS — DEVILS

There is never a shortage of positive guidance entering your world. It may seem like you are being inundated by the negative, but that is only because people choose to follow the negative and amplify it. They do this because they are wounded and their wounds fill them with pain, hatred, anger, jealousy, and sorrow. Therefore, they resonate to the negative.

They are wounded from events in this life as well as previous lives, but they also are wounded and trapped in their wounds by false beliefs and misunderstandings about life. They live in self-hatred and lack of acceptance of themselves and their mistakes. They believe themselves to be flawed, imperfect, bad, and doomed to unhappiness. They do not realize who they really are—God clothed in flesh, experimenting with a human body and feelings. They do not appreciate that their mistakes and struggles gain them strength and wisdom.

Being human is difficult even for those who are very evolved. Star People feel frustrated with being human because on some level, they know they are not human. And being human is difficult even in the best of circumstances and, heaven knows, very few are experiencing ideal lives. Many believe they should, however, and this is responsible for some of their pain. Expecting life to be happy is a false belief and like all false beliefs, it creates pain.

We are not suggesting that you not expect to be happy (better that than expecting to be unhappy!), but expecting to feel happy all the time is unrealistic. Happiness is a state that is attained when the ego has gotten what it wants, and it lasts until the ego decides it wants something else it doesn't have. Therefore, happiness is always transitory. It can be no other way.

Joy, on the other hand, comes from experiencing your Being, which is not dependent on circumstances. Joy is attainable and attained eventually by all as a result of evolution. Joy is the goal, not happiness. As long as you focus on happiness as a goal, you will be disappointed. You are not here to be happy. You are here to learn, to grow, to evolve, to love, to serve—and to experience the joy of your Being! Life was never meant to be easy. Reclaiming your spiritual Self is not easy, and you are not to blame for that. It is not your fault.

Planet Earth is a school. With every choice you make, you learn. That is all that is required of you—learning, not perfection. Being able to accept your mistakes and the mistakes of others will help you heal your wounds. What you believe about life and about yourself makes a big difference in how easily you can overcome the tribulations of life and the wounds. No one escapes being wounded; it is impossible. The lesson is in making the best of those wounds and learning to overcome them so that they don't block your progress. And sharing with others how you overcome your wounds can be a great benefit to them.

Life sets troubles in your path to test you and make you stronger. You can use them for that or you can wallow in anger, hatred, self-pity, and pain. Always, it is your choice. Since these feelings are so painful, people eventually develop the skills for moving through their difficulties. Nevertheless, these states can be enticing. What payoff do anger, hatred, self-pity, and sadness have for you? Do you get to be self-righteous? Do you get to blame someone else? Do you get to have an excuse for not doing something? Do you get attention from others? <u>Becoming conscious means becoming aware of not only your feelings but of the choices you make in regard to them.</u> How are you using your feelings?

When you ask us, we will help you by working with your emotional bodies. We have the means to speed up healing on all levels, but you must ask for it. Many of you are able to feel us at work on an energetic level, such as during your meditation or when you are very still. But much of this work happens while you are asleep because that is when you are the most receptive. Your conscious mind is out of the way and the unconscious mind can therefore be more effectively programmed.

If you remain in a negative state and don't ask for help, you are more vulnerable to negative forces and they, too, can affect you when you sleep. In a sense, because of resonance, your negative state gives them permission to interfere with you. You draw them to you by resonating at their level. The longer you stay in a state of negativity, the more vulnerable you are to negative forces. They can only make so much headway in the consciousness of someone who works to get and stay clear emotionally, such as someone who meditates daily.

You might think, then, that the negative forces focus on those who are negatively oriented. Well, they do, but they are even more interested in those working for the Light. They watch and wait for opportunities to affect them. That is why it is important for those of you who are working in service to the planet to guard your thoughts and work with your emotions. One of the traps is becoming lax about this. Most of the time you feel aligned with the Light, so you let down your guard, only to find yourself deeply depressed or discouraged. These feelings indicate that service-to-self is at work, undermining you.

When you find yourself depressed or discouraged, ask for help. It is also useful to take a moment to examine your thinking. Note the beliefs that are behind your discouragement. Then dismiss them one by one. You will find the beliefs that lie behind discouragement to be false. Here are some examples of the types of phrases that negative forces plant, so watch out for them:

- Nothing I do ever matters. What's the use?
- I will never succeed.
- I'm not getting anywhere.
- Nothing ever goes right.
- Everything is going wrong.
- Nobody cares. Nobody appreciates me. Nobody loves me.

Notice what these statements have in common: they are sweeping generalizations. Let the words "never," "nothing," "everything," and "nobody" be red flags indicating insidious lies that feed negative feelings and undermine positive action. That is just what the negative forces hope to do. But you can take charge simply by monitoring your thoughts and casting out the ones that don't belong to you. Your brain is a computer that is being used not only by you but by many others both in bodies and not in bodies. Many of you have learned to identify and correct the negative conditioning you received by your parents and others while growing up. Be as diligent about the programming you are now receiving by negative forces in your environment and in unseen realms. Knowledge is power. Awareness is your best defense. And again, we await your call for assistance.

Our main means of contact with you now is during sleep where we are preparing you for a more conscious meeting in the future and healing you to help you realize your Self and your mission. We are also showing ourselves to you, giving you tours of our vehicles and home planets, and telling you about ourselves while you sleep. Then, when you hear and see these same things in your waking life, you will not be so surprised; there will be a certain familiarity about us, and what we tell you will seem to make sense. Some of you even remember some of these encounters. However, in most cases, it is not our intent to have you remember them. We are intentionally working with your unconscious minds first.

We do have plans to contact you physically in the near future. We don't know when this will be because many factors are involved, mostly to do with your readiness, but also with the actions of the negative extraterrestrials. The sooner you are ready, the sooner we will present ourselves. The actions of the negatives could cause us to meet before you are ready. That would be unfortunate. The result would be more chaos than we would want to create, but this would only happen if the negatives were creating even more chaos. We hope you can appreciate why we can't be more specific about how or when we will introduce ourselves to you publicly.

We do want you to understand that we work cautiously behind the scenes for the most part and that we will intervene only if necessary. We respect your right to make the choices you are making on Earth in as much as they do not affect the welfare of the planet. But we are also here because your choices could have an impact on the well-being of those on other planets. So we will intervene to curtail any choices that might damage this or any other planets. In this sense, we are like good parents who allow their children freedom until they endanger themselves or others. We want you to understand our role on Earth and our deepest respect for your sovereignty.

The only other instance we will intervene is when you ask us to. Those of you who know of us can ask us to intervene on your behalf simply by making that statement. We can hear your requests and will fulfill them according to the highest good of all. Understand that the negatives will answer requests, too, but in fulfilling them, they will not do you the same honor of respecting your welfare and that of others.

Ensuring that your requests are met by us and not by the negatives is simple: address us specifically, either by name or more generally by referring to those who work in the Light. If you like, you can call on the Ascended Masters, angels, or other divine beings by name to answer your requests. Just as you would not mail a letter without a name and address, be sure to address your requests as a safeguard. If your motives are pure, your requests are automatically handled by those working in the Light. It's when you make a request out of anger, revenge, hatred, or the like that the negative forces may come forth to do your bidding.

When a request is made of us, we will see if fulfilling it suits your soul's plan and the plans of anyone else it might affect. Many people request contact with us, but few would find it anything but disruptive to their lives. It's not to anyone's benefit to have lives disrupted this way. That is why we have been so careful about appearing to you and erasing memories when we can. Many of you believe you could handle contact with us, but you would find it more upsetting than you realize. We know this and contact very few people in third density. Even our fourth density contacts are difficult when remembered.

It will be a while before we make contact with you in great numbers in third density. As a precursor, we will contact more individuals in third density and use them to help prepare others for our presence. This is already happening but will happen much more frequently in the next few years. Then, early in the next century, the general public will come to know us. It will not be long now.

The sightings have been stepped up and will continue to increase until the majority accepts the reality of extraterrestrial life, for at this stage it is still just a concept. More pictures and videos will be published to serve as proof. Crop circles, which we create, will proliferate and increase in complexity, bringing more of those with a scientific bent into the fold. As evidence builds, the media will contribute with more stories, movies, and docu-dramas on the subject to further prepare the way. Another important step will be the release of the government's classified information on this subject. Once this happens, it won't be long before you see us.

ET CONTACT

Chapter 7

Life with ETs

Building a New World Together

You will not be alone in the future. You never really have been. You will soon discover that you are part of a galactic federation. Many of those you will meet have been working with you all along without your being aware of it. Now the time has come for you to become aware of our presence and to co-create this universe with us. It is always a momentous time in the evolution of a planet when those inhabiting it first realize they have relations in other parts of the galaxy. You not only have relations, but they have known you and have been involved with you since your inception. You are not orphaned children of the Creator; you just haven't known your creators until now.

You are God-created, of course, just as we are; but your creation came about because certain beings chose to create your particular form using genetic engineering. This story has been told elsewhere, so we won't repeat it. Now you have reached adulthood and are ready to vote! First you will learn about the cosmic government and who's in it. That is part of what we will teach you. Our primary mission is to prepare you for what lies ahead. We will co-create your world with you and teach you to be co-creators of other worlds. Soon we will begin to work closely with you—as closely as you will allow—to build a new world.

We have been observing what you have created on this world. We have seen where your choices have led and are ready to step in with advice and guidance about how to improve what you have already created. We will also offer a vision of the future that will help you bring forth creations of which you have yet to dream.

Many of your problems have been caused by shortsightedness and ignorance of who you are, of your future. Once you realize your immensity and your latent powers, you will begin to make wiser choices about how you spend your energy. When you realize what power you have to create a positive reality, you will stop putting your energy into creating a negative reality. So much of your suffering is caused by simply being unconscious. You are becoming conscious now—conscious creators of your reality—and we are here to help guide you in creating the world of your dreams.

Please note that we said "the world of *your* dreams," not "*our* dreams." That is the difference between those who serve you and those who serve themselves. The world we wish to help you create will come from you, your desires and your goals. We will simply help you manifest them. We have no intention of imposing our perceptions and goals on you; but like a good parent, we will lead you to what we believe is in your best interest, while respecting your right to choose not to follow our suggestions. When we say we will build a world with you, we mean that we will work with you cooperatively, respectfully, and patiently, according to your needs and desires.

Is there any question that a new world must be built? It must be apparent that change is imperative. You cannot continue on your current course. Usually a world follows a course until it can go no farther, until it collapses. It does not necessarily have the wisdom to stop in midstream and change course. Many among you are wise enough to know that change is necessary now. With your cooperation, we will show you how to adjust your course before it's too late. You are headed toward ecological disaster. If you let us, we will help you avert that. We will convince you that you need to change your actions. Then we will help you correct your mistakes and find new ways to accomplish what you need to be comfortable.

Many of you are afraid to look at your world's problems; you have enough of your own, it seems. But if you keep hiding your heads in the sand, you will pursue your current path until you can go no further. We hope to wake you up to the urgency of your situation so that you can begin to move forward on a less destructive path. Nothing short of a shock may suffice since, for most of you, the drive to continue as you are doing is so great. The shock of having extraterrestrials enter your world may be just what is needed to wake you from your lethargy. So, although we are waiting to introduce ourselves until it's not too much of a shock, we will not wait so long that it becomes too late.

Life With ETs

What will your world be like with extraterrestrials living among you? How will that affect your societies, how you think, and what you believe? Because of our advanced intelligence, our broader perspective, and our lack of conditioning to your world, we will bring change to every aspect of society. Contact with us will shake many of your beliefs, release you from much of your conditioning, and change many of your culturally-determined values, including how you spend your time and how you see yourselves and life. As we present you with new possibilities, you also will change how you do things.

Some time early in the next century, we will live and work among you. We can and will be integrated into your societies, just like other cultural groups have in the past. We may not necessarily own homes or settle here, however, since few of us are adapted to your environment.

The first extraterrestrials involved with your world will live in spaceships and commute to your world to perform their tasks, although some may set up communities on Earth in special environments that allows them to be here full time. They will keep to themselves at first while you acclimate yourselves to them, coming into contact with you only when necessary to perform their work. At first, only a few will work among you, but in time more will come. Integrating ETs into your world will take time and require care and sensitivity regarding our impact on your world.

The Challenges of Integrating ETs into Your World

Integrating us into your society presents some unique challenges that other cultural groups have not faced. Language, however, will not be one of those challenges. We can readily learn any of your languages, and we also have the technology to help us communicate with you. We communicate primarily through telepathy and can induce telepathy in you. As in the *Star Trek* TV series, we also have the technology for translating one language to another. Many of us have studied your languages and will make an effort to communicate with you in the normal way so that you will feel more comfortable with us.

One of the biggest challenges will be the difference in our level of intelligence. It will not be as difficult for us to adapt to your more limited intelligence as it will be for you to understand us and appreciate our point of view. For instance, it will be difficult for you to evaluate some of our proposals and suggestions because they are so much more advanced than anything you can come up with yourselves. This will cause problems among the intelligentsia, who are bound to feel threatened by this influx of new intelligence into your world. They will have to learn to cooperate with and often take subordinate positions to us because of our greater knowledge and expertise.

On the other hand, not all the extraterrestrials you will meet will be skilled at adapting to your level of understanding. Some will not be able to put themselves in your shoes and see things as you do. Therefore, their advice may not be valid or it may not be understood by you. When misunderstandings occur, they will not always be your fault, but may occur because the extraterrestrial dealing with you lacks understanding of you or the skill to communicate at your level. Thus, intelligence alone is not enough to ensure good communication; an understanding of your cultures and what it is to be a human on Earth at this point in history is essential in dealing with you effectively. In addition, of course, there must be the drive to serve your best interest, and not all extraterrestrials are so motivated.

Another challenge will be adapting us into your work world. We will not always fit into your current structures because to us these structures are antiquated and highly inefficient. They will have to change. Your work is already becoming increasingly automated through technology and computers. This trend will continue as you find more ways to accomplish tasks without leaving the comfort of your homes.

We will help you create new life-styles around work by helping you develop your communications technology further so that centralized workplaces will be replaced by more conveniently located work spaces. This will change the pattern and routine of your daily lives and free many of you from your nine-to-five schedules. Your life-styles will change as you not only experience more freedom around your work activities but more free time and easier access to mentally stimulating and educational activities. As it stands, your free time is often consumed by mindless entertainment, which wastes human resources. You will be both entertained and enlightened in the future through the wise use of technology.

Education, knowledge, and expertise will be high priorities in the world of the future in part because the level of intelligence of the extraterrestrials living among you will demand more from you intellectually. It is not so much that you will be competing with us for jobs but that the technology we introduce to your world will demand a high level of education in order to use it. Even the average person will need what you now consider exceptional skills just to keep pace with the growth and changes occurring in society.

It will be difficult for those who can't keep pace intellectually but because technology will bring more prosperity to society in general, it will be able to assist or carry those who are least able to manage. The technology will help these individuals fulfill as much of their intellectual potential as possible. Also, because society will be less stressed and run more smoothly, there will be more time for spiritual development, and greater understanding and compassion for those with lesser abilities.

Another challenge facing your world concerns the sharing of what we have to offer. You will have to learn to share the information we bring if you are not to fall into conflicts over the rights to our ideas and our help. We are very aware of our potential to cause disputes among you and will try not to play favorites. We will work fairly with all people and foster cooperation between nations, not competition. We will work hard to bring your world together under one organization and government because this is the next step in your evolution and necessary for peace. Many of our suggestions will need to be implemented worldwide and will therefore require a certain restructuring of your political and social boundaries.

The Impact of ETs on Your World

Of course, the presence of extraterrestrials on your world will have a profound impact on your religions and belief systems. Our presence will cause you to question many of your cherished beliefs and ideas, including what you believe about human history. Your beliefs about your world and your history affect the kind of world you create. When you change your beliefs about your world, your world changes. After all, your attitudes, beliefs, and thoughts are what create your reality.

The new information we will bring will revolutionize many areas of study, such as geology, biology, archeology, cosmology, history, medicine, and mathematics. Opening your minds will trigger a major leap in consciousness and knowledge, since certain regressive ideas have thwarted your progress and limited your understanding. Once these ideas fall away, you will move ahead very rapidly. Extraterrestrials will advance your knowledge in these fields by hundreds of years in just a decade.

As a result of the medical advances we will bring, how long you live and how you pattern your lives will change. People will live longer and stay productive longer. There will be fewer debilitating illnesses and a healthier population as a whole. Consequently, there will be more free time, less difficulty in providing for yourselves, and a healthier life-style. On the other hand, there will be a greater need for advanced education and specialized training and therefore a much longer period of schooling before entering the work force. People will remain in the work force longer, but job-sharing will replace the full-time work that is now the norm. Therefore, even though the number of available workers will increase, unemployment won't.

These changes in turn will reflect a change of philosophy. When people have more time to devote to themselves, interest in life-enhancing subjects will increase. People will explore new pastimes as well as more serious concerns, such as the meaning of life. The struggle for survival and the emphasis on materialism so prevalent now in your culture will be superseded by more meaningful goals, such as personal growth and the attainment of peace and contentment. When you are able to meet your material needs more easily, you will not look so much to work and consumerism to satisfy your souls (which has never worked anyway). Thus, there will be an expansion of consciousness resulting in greater happiness. Going to war with each other will become a thing of the past.

The World of the Future

You are headed toward a Golden Age! It may come as soon as the next century or shortly after that. Meanwhile, you have a mess to clean up here on Earth before you are solidly on course toward a better world. You will suffer some from the consequences of your past choices that have damaged your environment, which will take time to correct. We will help you if you are willing to implement the changes we suggest. However, some of these changes are radical and will challenge some of you.

The biggest change, besides altering your personal habits around how you use your natural resources, involves establishing a worldwide agency for regulating and monitoring environmental issues. This will help bring you together as a world, although the logistics of doing this are challenging. Any global initiative such as this requires an infrastructure that is yet to be conceived, although a breakdown of national boundaries is necessary anyway for the equitable distribution of goods—a key to establishing peace on Earth. Global peace is foremost on our agenda, and we will try to convince you of the importance of the redistribution of wealth to peace. Not only will you find that peace will be a worthy goal, many of the benefits we bestow on you will be contingent on your willingness to strive for peace.

The world of the future will be very different. One of the biggest changes will be that the population of Earth will be much less than it is now. Natural disasters, widespread diseases, and increasing sterility will dramatically decrease your population in the near future. This, however, will make it easier for you to deal with the environmental problems and provide a healthy life-style for those who are here. Unfortunately, it often takes problems like this to advance a race spiritually.

Your world will have gone through other changes that will help you be more at ease with us when we finally do make our home here. For example, you will have made some breakthroughs in science that will help explain the paranormal. You are on the verge of a paradigm shift in physics right now that will open up your world view and your understanding of other dimensions and their inhabitants. You as a society will finally accept that sentient life exists in other dimensions and that these dimensions are just as real as yours. This will lead to an eagerness to learn from those in these other dimensions.

This will be a very important shift, because as it stands, science's skepticism not only inhibits belief in extraterrestrials but makes it necessary for us to stay behind the scenes. We await a shift in your belief system and are pushing for this by showing our craft to you. We know that moving too quickly will only cause problems in relating to us. A number of Star People working in the area of physics are striving to make the necessary breakthroughs for this next step. Through Star People, we are already having a far greater impact on the transformation of your beliefs than you realize.

Some of the things that will happen in your world in the next two decades will scare you and cause you to be more open to new ideas, new beliefs, and new solutions. Many of these will be supplied by Star People and other advanced individuals who have incarnated to help the planet solve her problems. Crises create an openness, which often results in leaps of consciousness and the transformation of structures that cannot be accomplished any other way, or at least not as rapidly. This will be the reason for many of the crises you will experience in the coming decades. They will challenge you and therefore strengthen you spiritually, emotionally, and intellectually. The universe has a beautiful way of providing for our needs by delivering challenges. As participants in Earth's plan for transformation, we can offer some of the help you need to advance spiritually and intellectually.

Not all the extraterrestrials you will become involved with will serve you well, however. You need to be aware that there are many races with a variety of motives, some of them self-serving. One of your lessons will be discerning who to trust and who not to. Meanwhile, you will probably make many mistakes and learn from them, but that, too, is part of the plan. This is one of the ways you will advance spiritually and become wiser and more aware of the importance of adhering to certain values. The unscrupulousness of some extraterrestrials will result in debates about what is right and wrong, moral and immoral. In the future, having people in power who have integrity will be especially important since the potential for destruction and the abuse of power will increase as a result of your intellectual and technological advancement.

You will also experience improvements in health as a result of our presence among you. Your world will be much healthier, not only because the environment will be cleaner but because you will develop technology to overcome many of your illnesses and diseases. You will also have a better understanding of the basis of good health.

Many of the advances will be accomplished through genetic engineering as we teach you how to unplug, so to speak, the genes in human DNA responsible for unwanted health conditions. This will eliminate much suffering on Earth.

We have other technology to share that will ease suffering and improve lives. Advances in transportation, for instance, will make it possible not only to travel swiftly and safely anywhere around the globe but to also travel among the stars. These advances in space travel are probably more than a century away but that is still a very short time to go from cars and planes to interdimensional travel. Even though we can already do this, you will have to develop the technology on your own to some extent. For us to just give you this technology would be too much of an interference; moving too quickly into space would be too shocking for humanity. It would also deny you the opportunity for intellectual development that learning to travel through space affords. Therefore, you will be given only the help that will benefit you.

We are very aware of our potentially enormous impact on you. We will be careful how we interact with you, and with the technology and help that we give you. Some of us have made mistakes before with other races—some even with your race—and have learned from these mistakes. Although they also operate under a certain ethical code, other self-serving extraterrestrials will not be so careful, not because they are less wise but because they hope to throw you off course. They hope to cause problems on Earth because creating chaos is how they gain control.

Self-servers have always been part of your universe. There are even whole planets inhabited primarily by self-servers. Whenever they have the capability of traveling through space, they try to disrupt and conquer other civilizations. Sometimes they are successful and sometimes not; much depends on the orientation and development of the targeted civilization. The less evolved and more oriented to self-service a civilization is, the easier it is to conquer. That is why we are not greatly concerned about the fate of Earth. As unevolved and self-serving as humanity may seem at times, you are far better off than many other races. Because self-servers are a relatively small percentage of your population and because many highly evolved individuals are here to help raise the mass consciousness, the best the negative extraterrestrials can hope to accomplish is to create some chaos, fear, and trouble, although they hope to do more than that.

You are moving into a very special time, one of great change and innovation. Many of society's structures will have to adapt or change entirely to fit the future level of consciousness of Earth's inhabitants. Some of these are already here and have managed to raise their consciousness through effort and grace. Others will reincarnate here because the level of consciousness suits them and their growth. There will still be a range of consciousness on Earth—not everyone will be at the same level—but the range will be narrower than it is now and the average level of consciousness higher. That is good news not only for the people here but for Earth itself, for it will need careful stewardship in the future to recover from past sins.

It may seem like you will never solve your current problems, but many of them are simply growing pains more than anything else—signals for the need to change. Once certain changes are made, these problems will be resolved. Of course, there will be new and different challenges, but you will have more resources to deal with them than you do now. Challenges still exist in more evolved states, but the pain associated in dealing with them is much less. Growth continues but through less pain and with greater swiftness. A higher level of consciousness brings with it an ability to master thoughts and emotions, which are behind the creation of your reality. The more evolved you are, the better creators you are and the more you are able to change what you don't like or what doesn't work.

You are going to make giant strides in understanding how the brain, mind, and consciousness work. Because you will realize that all else comes from this, this will be the new frontier. Solutions to your problems lie in understanding the mind and consciousness. Mastery of the mind and the effect of mind and emotions on the electrical, chemical, and hormonal activity of the brain will be crucial areas of study in the near future. You have already begun these studies, but so much more will be uncovered. We are well-equipped to help you with this. Many abilities that you now consider psychic will unfold as you master your minds and enlarge your brain capacities. Although many of us have brains unlike yours, some of us have brains that function as yours will in the future, so we can help your scientists unlock those secrets.

You will begin to understand the nature of reality from the standpoint of dimensions where time and space do not exist. For this, your current science is useless. A new science will evolve to account for other dimensions and you will realize your multidimensionality. Of course, this will have enormous spiritual repercussions; however, science will transform religious and spiritual views, and not replace them as it has in the past. Your whole concept of reality will soon change as a result of meeting us since we already live in and travel between other dimensions. Our knowledge is your future science. You have so much to learn from us.

We have much to gain from you, too, primarily in service to you. We are here to serve you. Some of us are here to make amends to you for past mistakes and others to make amends for mistakes made elsewhere. We need you as much as you need us. We are as motivated to help you as you will be to learn from us.

You will learn from both the positive and negative extraterrestrials, although what you learn from each will be very different! The negatives will teach you discrimination and the value of honesty and integrity by demonstrating the opposite. They will show you what happens when a race falls out of alignment with Spirit and they will play out the negative for you so that you can integrate your own shadow side. They are performing an important function in your world in bringing you certain lessons. For this, you can be grateful, just as you are grateful for other challenges in your lives that provide fuel for your evolution.

Change accelerates evolution. The introduction of extraterrestrials into your world will accelerate your evolution exponentially, but you will have to be willing to see things differently. Flexibility and adaptability will be essential as you adjust your beliefs and what you know about the world to accommodate a much broader reality. Life as you know it will no longer exist, not only because the structures of your lives will change but because the structures of your brains, your beliefs, and your very perceptions will change. You will change so fundamentally inside that your outer reality cannot help but change dramatically. You will not be in Kansas anymore!

This is how the new world, the Golden Age, will come about. It will happen as a result of deep changes in your brains, in how you think, in what you believe, and in how you perceive. Reality, after all, has always reflected consciousness. Therefore, it should be no surprise that the greatest leap in consciousness that humanity has ever taken will result in a new world, one that will not only look different but will serve your evolution differently. This new world will suit a certain level of consciousness and serve certain evolutionary lessons. Those still learning the lessons of the old Earth will not be doing it here anymore. They will complete these lessons elsewhere. As Earth moves into a new consciousness, a new dimension, it will serve the Creator in a new way.

Chapter 8

What You Can Do to Help

It must be obvious by now that you are very important to our plan, which is to say that you are very important to Earth's plan. We need you to carry out tasks in third density that we cannot. And we need your permission to help you. We cannot help you unless you want our help and ask for it. This is your world; we are just here to help you move it into the next dimension. We are your guides, but you are the creators of your reality. At best, we can be co-creators with you.

In this chapter, we would like to offer some practical suggestions about what you can do to help bring your planet and yourselves into fourth density.

What You Can Do to Help Yourself

Each of you is important. Each of you is unique. No one else is exactly like you, not only on this planet but anywhere in creation! What you have to contribute, no one else can contribute. And each of you has something to contribute, even those apparently without valuable skills. You are so much more than you realize!

Each of you has the potential to bring Light to this planet in a number of ways, not just with your talents. You don't have to be talented or educated to be a Lightworker, or to serve. Some of your life tasks require specialized skills or psychic abilities or other talents developed in previous lifetimes on other planets and in other dimensions. But please appreciate that Spirit can and will use even the humblest and least developed of you if you are willing. You do not need to be evolved, intelligent, charismatic, beautiful, talented, a Star Person, or have any other quality except a willingness to serve. Of course, that usually does entail a certain level of evolution.

The most important thing you can do to help is to offer yourself in service to the planet in any way Spirit sees fit to use you. Do not put strings on it such as, "I'll be of service as long as it pays my bills." And do not judge the form this service may take. The ego is so quick to compare and judge. No one kind of service is more valuable than another. Some forms are certainly more prestigious, but these judgments are those of the ego, not Spirit. Whatever service is needed is needed, and therefore valuable.

What's more, you depend on each other's services to fulfill your own. For example, a teacher cannot teach if someone doesn't print the books. So who's to say a teacher is more important than a printer or a machinist who makes the printing presses? You do not have to be a psychic, a workshop leader, or a healer to have a role in the planet's transformation.

Besides, what you do is only part of the picture of your service to Earth. *How* you serve is equally important. Do you spread love with words and actions throughout your day? Love is the stuff that planetary transformation is made of. Living in loving kindness is the most important thing you can do for the planet and for humanity now. This is something anyone can do. You don't need a degree to love and it doesn't cost you anything. In fact, it pays very well! *There are no excuses not to love. Living in love is a choice you make every second of your life.* You are continually choosing love or separation. What is it that stands in the way of your choosing love? The answer to this question is different for every person.

What you can do to help is to love and to heal all those parts of yourself that prevent you from loving continually. Of course, this is a tall order, but you don't have to do it perfectly. Just do the best you can to keep choosing love until it becomes a way of life. You will eventually become very good at it! Because loving is its own reward, it becomes easier and easier each time you choose it. You may need some help at first to heal whatever stands in the way of making that choice, but that is what healers are for. Plenty of help is available to help you heal the wounds that keep you from having an open heart.

What is being asked of individuals and humanity as a whole now is to open the heart. The raising of consciousness occurring on your planet is not an intellectual process (although your brain capacity and usage are expanding), but a deep emotional clearing, which will result in greater love and compassion. What you can do for yourself to improve your ability to serve is to heal your heart. When you do this to help yourself, you will be helping everyone else.

Those who will be living on Earth in the future will have open hearts. This will change the form that your lives will take, along with every structure you have on Earth. When people begin to operate from their hearts, you will live in a very different world. When we talk about moving into fourth density, we are talking about living in a world where love prevails instead of fear, hatred, and anger. But in order to have a world like this, you each must do the clearing necessary to allow for the opening of your heart.

Healing is in order now. It is the most important thing you can do for yourself and for the planet. There is no getting around it, and no one can do it for you. However, we don't want to leave you with the impression that you don't have help with this. Spiritual forces are working with each of you daily to help you clear your emotional bodies in preparation for a new world. Your part is to open to these forces, invoke them consciously, and ask them to help you become a clearer instrument for service.

For some of you, this will require dropping whatever false beliefs interfere with taking this step. What stands in the way of you making this request for healing from Spirit? Are you having a hard time setting aside time for meditation and prayer? Do you feel you aren't worthy to be an instrument, to receive the help and love that is yours for the asking? Do you feel silly? Perhaps you don't believe such forces exist or can help. Or perhaps you don't feel you need healing. That is the arrogance of the ego. The ego will try to interfere with this process. The challenge is to overcome the ego's ploys such as telling you that you don't need any healing, that you're okay. Observe all the other ways the ego may try to keep you from making a commitment to your own healing or to meditation.

Your ego is there for your survival. It is a primitive aspect of yourself that keeps you safe through fear, but it has a narrow view of safety and therefore feels threatened by change or anything that doesn't immediately relate to survival. Although the motives differ, the negative forces use exactly the same ploys that the ego uses. Therefore, distinguishing the two can be difficult. Both the ego and the negative forces use fear to keep you from moving forward.

Another trick the ego may use is to tell you that something so simple as asking can't possible work. "Why bother?" it says to you. "It can't possibly work. It can't be that simple. Besides, you have more important things to do, like earning a living." Is it so inconceivable that you should receive exactly what you need just by asking for it? You see, life doesn't have to be as difficult as it is for many of you. You don't have to go it alone. We are just waiting to help out, but you have to show us you want help. Some of you enjoy your pain too much, and we can't help you until you really want out of it. That might sound odd, but it is true. Many of you get something out of your pain that you are unwilling to give up. And others don't ask for help because they feel they deserve their pain and are unworthy of something better. All you have to do is ask. It's so simple yet so hard for some.

It only takes a moment to ask. The more heartfelt your statement, the more help you call forth. So asking is important; but did you realize that by increasing your intent, your desire for help, you increase the response? How much of your will do you put behind your request? Just a little or your entire being? That tells us just how much you mean business and therefore how much we should help. We will help as much or as little as you ask us to. It is up to you.

Notice what may be keeping you from putting your full intent behind a request for help. Don't just notice what keeps you from asking in the first place, but what keeps you from asking wholeheartedly? You will come up with some different answers to this second question. Do you really want healing with your whole being? Do you really want a relationship with your whole being? Do you really want to discover your life's work with your whole being? What parts of yourself don't want those things? Those are the parts that keep you from asking fully and receiving fully.

What You Can Do to Help the Planet

Whatever you do for the planet, you will be doing for yourself, of course. You and the planet are inseparable and interdependent. That means that not only do you need the planet but the planet needs you. That may sound strange. Why does the planet need human beings? The higher life forms on the planet are just as much a part of her as are the so-called "lower life forms," the minerals and plants.

You are simply a higher expression of the planet's energy. You are the planet. If the planet fails to sustain you, she has failed to sustain herself, and she feels this loss.

The planet is working very hard to sustain the growing human population, but you must also help by realizing the strain you put on her and reduce it. She can only accommodate so many of you. Her capabilities are not infinite. To you, they seem unlimited but you are testing her to the limits now.

What you can do to help the planet is to cut back on pollutants, replant the trees you take, clean up her oceans and other toxic sites, and stop producing nuclear and other toxic waste. The toxins you produce throw off the balance of Earth, making it harder for her to support you. The fish are dying. What will you do if the oceans die? The answer is that you would be dead long before the oceans reach that point, so you can't wait until then to do something about it.

There are already many organizations in place that serve planet Earth, so one obvious way you can help is to support these organizations. Less obviously, you can pray for the planet. Just as asking for healing for yourself invites help from other dimensions, so does asking for healing for the planet. You use your free will to damage the planet, so why not use it to correct these problems through prayer and other positive actions?

Again, unless you ask for our help in this matter, we can only intervene to a certain point. We are limited in what we are allowed to do without your permission. We can only go so far. Much of what we are doing now is research and observation so that we are better prepared to intervene once your governments ask us.

What You Can Do to Help Us

Many of you are already helping us with all our various projects without knowing it. However, there is always need for more help, so if you have not already signed up, then please do so simply by stating to us in your meditation that you would like to assist those working in the Light, specifically those working to serve whatever course you wish to name. It is as simple as that. Then, if your request fits your talents and your soul's direction and plan for this lifetime, it will be done. Chances are, however, you are already in service in exactly the capacity you would like to be!

You are far more intuitive than you think. What you love to do, you are already doing in other dimensions. This enthusiasm is a sign that it is already being done elsewhere, or at the very least, being prepared for. In this dimension, you feel limited—you judge yourself as not smart enough, not strong enough, not talented enough, not _____ (fill in the blank) to do what you want to do. Those of you who are more practical limit yourselves by telling yourselves that what you want to do will not pay the bills. In other dimensions, you do not limit yourself in any of these ways. You simply do what brings you joy. That is what you must do in this dimension as well. Don't let anything limit you. The secret of being happy on this dimension or any other is to do what brings you joy! How much less pain there would be in your world if everyone followed their bliss!

What keeps you from following your bliss? Examine this carefully, because the answer is what the negative forces use to inhibit you from fulfilling your mission. If the answer is that you're afraid of being poor, the negatives will tell you that you won't be able to survive if you do what you love. They may even inspire others through their unconscious minds to tell you that. If it is fear of rejection by your parents or other loved ones, the negatives will tell you that this is what will happen if you do what you love, and they may even encourage your loved ones subconsciously to reject you.

But you are here to live your life as you feel called to by Spirit. Not doing so will bring you pain and waste your special skills. You are not called to do something for which you have no skills; that would not make sense. Your soul is very sensible in planning your life. You can trust that what you love to do is something you have enough talent for. You may not be the best, but you don't have to be to make a significant contribution. It may be several lifetimes before you will use your talents optimally, but that is no reason not to use them now and take them as far as you can.

Trust yourself. Trust your feelings. They will tell you what you need to be happy and fulfilled. Don't judge your feelings. They are part of your humanity, a very important part, and they serve a purpose. No other race is exactly like human beings here on Earth. You are learning to understand and use your feelings positively. In order to do that, you must trust them. First, be aware of them, then acknowledge them, then accept them, then make your choices according to what you have learned from them. That is how you can find your way to your right form of service.

Your soul's plan is instigated through your feelings and intuition. Your guides and others like us use these to lead you to your life work. So while you do not need such guidance when you are in other dimensions, you do need them in this dimension. Many of you feel confused, lost, and uncertain of your path but you don't have to. If you feel this way, you are not trusting your feelings and intuition. You know exactly what you need; you just don't know that you know it because you don't trust that you do.

What is it that causes you to distrust your feelings? Well, many of you have discovered that following your feelings has gotten you into trouble. However, your feelings are not what get you into trouble, it's what you decide to do with them. Feelings tell you what you want. What you do after that to get what you want may be either a good idea or a bad one (or somewhere in between).

Feelings themselves are neither good nor bad, but sometimes they lead to negative consequences. Therefore, you don't trust. You throw the baby out with the bath water. That is one of the lessons having feelings teaches you. So the answer to the question, "What can you do for us?" is to trust your feelings and your intuition. By doing that, you will be in alignment with your plan; and when you are in alignment with your plan, you will be fulfilling your mission and therefore ours.

We don't expect you to perform your mission perfectly. Thinking you have to do something perfectly keeps some of you from ever beginning. We accept the imperfection of third density and the imperfection of being human. You, too, must learn to accept it. The work still gets done, even if imperfectly. Evolution continues to happen, and plenty fast at that! Sometimes your imperfections even speed it along by bringing you very dramatic lessons. So, please, follow your heart. You cannot go wrong, because going wrong in the process of following your heart will still get you there—maybe even faster!

Another thing you can do for us is to consciously align yourself with us. You do this simply by making a statement to that effect, out loud if you can. In so doing, you also help to fend off any negativity that might be trying to enter your consciousness. Every day, several times a day, affirm your desire and intent to be aligned with the Light and all forces working for the Light. By continually affirming this, you make it difficult for the negative forces to use you, and that helps us! Stay clear and positive emotionally and mentally, pray for love, peace, healing, and goodness on Earth. Trust your feelings and intuition.

Bless you and thank you!

CHAPTER 9

On the Way to Ascension

What Is Ascension?

Ascension, as we define it, means moving into another dimension and all that it entails. You and your planet are ascending into fourth density. Fourth density differs greatly from third, so the consciousness and any reality created from that consciousness will also greatly differ. Thus, fourth-density Earth will be a very different place from third-density Earth, and those who live there will also be very different.

Not everyone who is here today will stay or return to the planet in their next incarnation. Many will choose to be reborn on another third-density planet where they can continue with the lessons of third density. Those who no longer need third density will either remain here as Earth ascends or return in their next incarnation.

Some people currently on Earth are almost already in fourth density consciousness but cannot fully express it because the density of the planet and her population will not support such expression. This is not yet a fourth-density world! Planet Earth is quickly moving to this level, but the human population is still mostly in third density and this is causing some problems on your world. Over the next several years, however, many of these people will leave the planet and that will make it much easier for those remaining to move forward and express a higher level of consciousness.

Every planet and every human being eventually ascends, and today this is happening in greater numbers than ever before! Many people have incarnated at this time specifically to do just that. This is an excellent opportunity for many to move forward. In addition, the planet needs these individuals to make the necessary global changes for a higher state of consciousness. No one is here just for their own growth; always there is a give and take. Those benefiting in these important times serve the planet and vice versa. It is a privilege to be alive today on Earth. Many others who would like the chance have not been given it, because there are simply not enough available bodies. Those of you who are here are fortunate indeed!

Many of the extraterrestrials observing the planet are here because of this shift to fourth density. Some are here to help; some to observe, just as one might observe a loved one's graduation; and some are here to try to stop it. Ascension cannot be stopped, but some of the negative groups are trying to delay or derail the process. Every person whose ascension is delayed is a victory for the negative forces.

Ascension means being free from the suffering and conflict of third density. The negatives, however, wish to perpetuate the suffering of third density. When people leave third density, it proves that light overcomes darkness. That is not their reality and they don't want it to become yours.

Ascension fills you with Light. In the ascended state, you live in peace, love, and harmony with all. The negative forces believe that this state is a state of weakness. They do not see the strength of love, nor do they believe in it. They believe it exists, but they do not believe it is worth striving for. And some striving is required! Ascending takes a commitment to overcoming the ego, which looks out only for itself. Indulging the ego is the path of least resistance. It takes little effort to follow this path, although the consequences are often troublesome! Negative forces do not believe in self-mastery, which is overcoming the ego, but in mastery over others. They use their will not to conquer their own weaknesses but to exploit the weaknesses of others. That is their idea of mastery.

Anytime you work at overcoming negativity, hatred, sloth, discouragement, jealousy, and greed, you gain self-mastery. The negative forces use these negative states to control you, to master you. One reason life is so hard is that these forces work to make it so. Not only do you have your own human foibles to contend with but you also have the negative forces who exploit these foibles. However, once you know of the negatives' existence and how they operate, you are empowered to master them rather than be mastered by them.

There are fourth-density and even fifth-density worlds that have not ascended. These are the service-to-self worlds where only those dedicated to serving themselves live. Those in these worlds evolve intellectually, but not spiritually. When they finally choose to relinquish the path of service-to-self, they make an enormous leap in their evolution, however, then they also have to face the karma they created on these worlds, and balancing that may take eons. They do this in third density and then ascend to fourth density when they have completed their karma. From the Creator's standpoint, this is just another way to evolve, a more difficult one to be sure, but those who make this choice cannot be convinced of this.

Your world is moving into fourth density and you will need to ascend in order to remain here. That does not mean that third-density Earth will cease to exist. Third-density Earth will not disappear; it will continue in a parallel reality, in many parallel realities in fact. And many who are not yet ready for ascension will return to an Earth in one of those parallel realities. In these worlds, life will continue as if ascension never occurred. Those choosing this experience will learn important lessons about the choices they have made, now and earlier, that are not in alignment with ascension. It will not be easy.

To understand this better, we will say more about parallel realities. When an individual or a planet comes to a crossroads, every possible choice at that crossroads is explored in a parallel reality. Thus, there are many parallel versions of you living parallel variations of your life in which you explore the consequences of each choice you could have selected. One parallel reality is not more real than another, but an individual's consciousness resides in one reality more than the others, and that reality will be the real one to that person. This does not make any of the other realities less real.

So, when we say that Earth is ascending, we mean that many people on Earth today will experience ascension as their reality. They experience it because they are ready to experience it, meaning that they are willing to believe it is possible, willing to create it, and willing to receive it.

When we say that Earth is moving into fourth density, we speak to those of you who on some level wish to create this and are willing to become a part of this reality. Those who are not ready or willing to conceive of it as a reality will not experience it. They will continue in their third-density lives on Earth. Therefore, at some point, there will be a splitting off, and that is the momentous occasion that many of you are preparing for and awaiting.

Third-density Earth may or may not survive as a place for humanoid evolution. Much depends on the choices made. In one parallel reality, it could be taken over by negative extraterrestrials. In another, it might destroy itself through nuclear war. In yet another, it might turn the corner and ascend later. There are infinite possibilities and therefore infinite parallel realities connected not only with third-density Earth but also fourth-density Earth. Anything that is possible exists in some parallel reality. This is how the Creator explores all possibilities and learns from them.

You cannot fully grasp this at this time, so we and others like us try to explain this to you in various ways. We hope we have succeeded in clarifying this somewhat. We accept that you cannot completely understand it and you should too. Much of what is given to you through telepaths is difficult to put into language because our reality differs so much from your current reality.

We hope you understand that one reason the information coming through telepathically varies is because of the language problem. Not only do telepaths have different facilities with language, which influences the message, but your languages lack words for the concepts we try to convey. So we do the best we can with the available language. We are sure you can appreciate that a new reality will require new words to describe it. In trying to convey our concepts to you, we risk sounding contradictory and irrational, but please do not dismiss our message as a result.

The communication process is also open to inaccuracy and interference by others who are trying to mislead you. Nevertheless, it is the best and most direct means we have for communicating with you until we can be face-to-face with you. When that day comes, misunderstandings will still occur, but they won't be such a problem. Besides, everyone will then at least be listening to us. Right now, only a fraction of you are willing to listen or read material such as this. Not everyone is open-minded enough or up to the challenge of sifting through it.

This material challenges your ability to think for yourself and discriminate; not everyone wants that responsibility. Many would rather be told what to believe and what not to believe. Hopefully, those of you who read material such as this are willing to question your beliefs and make your own decisions about what you believe. There are some who accept anything without discernment, but they can't do that for long without running into contradictions. That's when many stop reading, finding it easier to toss it all aside than sort through it and decide what will be your truth. We appreciate those of you who have the courage and tenacity to strive for your truth.

No one will be able to explain ascension to you in a way that you can understand until you actually experience it. But hopefully, our explanation and those of others will help you prepare for it. We hope that just because the information coming through different sources is confusing and sometimes contradictory, you won't dismiss this important message. Fortunately, most of you take this information in through your heart and will find that it resonates there as truth even if it doesn't make much sense to your rational minds. Still, it can be frustrating to know and trust something without being able to put it into words or prove it to others.

Most of you feel uncomfortable about believing in ascension because it is a radical idea. However, those of you who do believe it don't seem to be able to help it! You just know it is true. But it will help you feel more comfortable and safe if you simply live your lives as you would if you did not hold this belief. This helps ensure that your ego or the negative forces don't use this belief to avoid the responsibilities and growth that are part of your lives now. It is tempting to conclude that you don't need to get an education or bother with other mundane duties if you are going to ascend. Beware of that kind of thinking. It is dangerous to change your future plans based on the belief that you and the world are going to ascend. You are still the creator of your reality, and you need to create the ascended reality and perform the tasks you came here to do!

Preparing for Ascension

What can you do to prepare for ascension? First, it is vital that you keep your emotions and mind free from negativity, which is why we have made such an effort to explain negativity and the negative forces to you. Become aware of your thoughts and feelings every moment and ask yourself where they are coming from. Then choose whether you will own that thought or feeling. If it is not helpful, discard it because it either doesn't belong to you or no longer needs to belong to you.

If negativity or difficult feelings persist or have an obsessive quality to them, find a healer who can help you. You may need to have an entity cleared or undergo past life therapy. Meditation may help, but many need both meditation and healing to get beyond deep issues or entity intrusions. And don't be afraid to ask for help; Lightworkers and Star People are bombarded with negativity these days.

Meditate. This is very important because it helps to clear your mind and emotions, and makes room for Spirit to communicate with you. Meditation is the most important thing you can do to prepare yourself for ascension. It clears your energy field and allows you to hold more Light. It literally enlightens you!

Prayer. We hope we have impressed upon you the importance of asking for help. Prayer is simply invoking the help of higher forces. Ask for strength, courage, faith, guidance, healing—whatever you need. You will receive exactly what you need in accordance with how much you want it and are willing to receive it.

Affirm your alignment with the forces of Light. You are on their team and they are on yours! Know it, affirm it, and declare it daily. Weed out anything that stands in the way of feeling aligned with the Light. Did you do something wrong for which you need to make amends and forgive yourself? Then make amends, forgive yourself, and move on. Do you feel you are not worthy enough? That is never true. Dismiss that thought, ask for help from higher forces, or seek help from a healer.

Open your heart to others. In serving others, even in small ways, you serve all. And in serving all, you serve the Creator. The beauty of this is that you also serve yourself because the rewards of service are so great. Service heals. It heals your heart and balances your karma. Find ways to serve every day. Any kind of service will do. It is the act of *choosing* to serve that is so powerful, not only to the one receiving but to the giver. When you give, you feel good about yourself. And feeling good about yourself makes it easier for you to receive the good that is yours. Giving can also help overcome guilt or the sense of unworthiness that can stand in the way of feeling aligned with the Light and receiving all that the Light has to offer you.

Be kind. When you are kind, you heal others. It makes it easier for them to act kindly in turn. Kindness snowballs: one act of kindness leads to many others. Unfortunately, unkindness works the same way, which makes it all the more important that you are kind, even in the face of unkindness. Today, why not make kindness your mission? It is a powerful choice, and one that will also heal you.

Do what makes your heart sing. You have a unique mission to accomplish, and you discover it by following your heart, by doing what you love. What keeps you from following your heart? Probably fear. Work through these fears. Get help from a healer if need be, but don't let fear stop you. Fear is no reason to be unfulfilled. If you won't do it for yourself, then do it for others because the rest of the world needs you to be doing what you came here to do!

Learn to receive and appreciate. So much in life is given without struggle. The beauty of this Earth, the love and friendship of others, the pleasure of your bodies and senses, the sweetness of a flower, the smile of a child—these are gifts free and available to all. Those of us who are no longer in a body, who can no longer participate in the joys of physical life, fondly remember these treasures. Be sure to let these joys touch you. Your journey will be so much easier.

And finally, *laugh!* Take yourself and life more lightly. The life you are living is just one of the stories in the book of who you are. If only you knew the whole story!

Peace to you!

ABOUT THE PUBLISHER AND LOGO

The name "Oughten" was revealed to the publisher in 1980, after three weeks of meditation and contemplation. The combined effect of the letters carries a vibratory signature, signifying humanity's ascension on a planetary level.

The logo represents a new world rising from its former condition. The planet ascends from the darker to the lighter. Our experience of a dark and mysterious universe becomes transmuted by our planet's rising consciousness — glorious and spiritual. The grace of God transmutes the dross of the past into gold, as we leave all behind and ascend into the millennium.

ABOUT THE ARTIST

David Adams has been painting since the age of twelve, first landscapes and, following the Harmonic Convergence, visionary art. David lives in the Pacific North West and works as a crystal carver and inspirational jeweler, yet still finds time for his art.

For additional information, please contact Oughten House.

PUBLISHER'S COMMENT

Our mission and purpose is to produce and disseminate ascension and higher consciousness information and materials for the enhancement of personal and planetary consciousness worldwide.

We currently serve over fifty authors, musicians. and artists, and we need your support to get their messages to all nations. All our financial proceeds are recycled into new inspirational books and expanding our distribution worldwide. If you are in a position to become an investor in this important work, Oughten House would welcome your contribution. Please contact us.

OUGHTEN HOUSE PUBLICATIONS

Our imprint includes books in a variety of fields and disciplines which emphasize our relationship to the rising planetary consciousness. Literature which relates to the ascension process, personal growth, and our relationship to extraterrestrials is our primary focus. The list that follows is only a sample of our current offerings. To obtain a complete catalog, contact us at the address shown at the back of this book.

Ascension Books

The Crystal Stair: A Guide to the Ascension, by Eric Klein. This is the book that put "ascension" on the map of human consciousness, worldwide. A collection of channeled teachings received from Lord Sananda (Jesus) and other Masters, describing the personal and planetary ascension process now actively occurring on our planet.
—ISBN 1-880666-06-5, $12.95

The Inner Door: Channeled Discourses from the Ascended Masters on Self-Mastery and Ascension, by Eric Klein. In these two volumes, intended as a sequel to *The Crystal Stair*, the Masters address the challenges of the journey to ascension.
—Volume One: ISBN 1-880666-03-0, $14.50
—Volume Two: ISBN 1-880666-16-2, $14.50

Jewels on the Path: Transformational Teachings of the Ascended Masters, by Eric Klein. In this book, the ideas and themes introduced in Klein's earlier books are clarified and refined. The reader is brought up to date on what exactly the ascension process consists of and how to be a more active participant in it. Current topics, such as the controversial Photon Belt, are also discussed. This is the best one yet! —ISBN 1-880666-48-0, $14.95

An Ascension Handbook, by Tony Stubbs. A practical presentation which picks up where *The Crystal Stair* leaves off and includes several exercises to help you integrate ascension into your daily life. Topics include energy and matter; divine expression; love, power, and truth; breaking old patterns; aligning with Spirit; and life after ascension. A best-seller! — ISBN 1-880666-08-1, $12.95

What Is Lightbody? Archangel Ariel, channeled by Tashira Tachiren. Articulates the twelve levels of the Lighbody process. Recommended in *An Ascension Handbook*, this book gives many invocations, procedures, and potions to assist us on our journey home. Related tapes available. —*ISBN 1-880666-25-1, $12.95*

Heart Initiation, by Julianne Everett. Answers many questions about the process of awakening: Does self-mastery have to be difficult? Why is love so important? How do we become truly free? What are the challenges and rewards of conscious ascension? Meant to assist you in surrendering as gracefully as possible to your own spirit. (Related tapes available.) —*ISBN 1-880666-36-7, $14.95*

My Ascension Journal, by Nicole Christine. Transform yourself' and your life by using the journaling methods given in this book. Includes several real-life examples from the author's own journals, plus blank pages on which to write your own ascension story. This quality hardbound edition will become a treasured keepsake to be read over and over again. — *ISBN 1-880666-18-9, $24.95*

Gifts: Remembering the Now, by Yolanda Zigarmi Martin. The author undergoes spontaneous past-life regressions, and interacts with several of her other incarnations. Spanning thousands of years, this process yields insights that reveal the very essence of our existence and our relationship with our Creator. A beautiful and thought-provoking book. —*ISBN 1-880666-59-6 $13.95*

Love and Hope: The Message for the New Millennium, by Kiyo Monro is a distillation of the wisdom and insights gained from a lifetime of dedication and service to ascension. Kiyo's style reveals the true humility of Lightwork and the gentle, loving words of the Masters encourage and inspire us, even when we might feel overwhelmed by the everyday world. This is a perfect "starter book" for those new to spirituality and the ascension movement.
—*ISBN 1-880666-56-1 $14.95*

Angels of the Rays, by Mary Johanna. This beautiful, full color book presents twelve different angels to assist you in your own healing and ascension process. Each angel has her own color, ray, gemstone, essence, invocation, and special message. These appear in the book and on removable Angel Cards for convenience. Makes a wonderful gift.
—*ISBN 1-880666-34-0 $19.95 Additional cardsets available for $12.95*

Tales and Teachings

The Extraterrestrial Vision: The ET Agendas — Past, Present, and Future, by Gina Lake. The non-physical entity, Theodore, tells us what we need to know about our extraterrestrial heritage and how to prepare for direct contact with those civilizations which will soon be appearing in our midst. Highly practical and timely information, given by a wise and loving teacher. Related tapes available.
— ISBN 1-880666-19-7, $13.95

Lady From Atlantis, by Robert V. Gerard. Shar Dae, the future empress of Atlantis, is suddenly transported onto a rain-soaked beach in modern-day America. There she meets her twin flame and discovers her mission: to warn the people of planet Earth to mend their ways before Mother Earth takes matters in her own hands. Be prepared for a surprise ending! — ISBN 1-880666-21-9, $12.95

The Corporate Mule: Giving Up Your Soul for the Company Goal, by Robert V. Gerard. In this slice of life book, follow a young man's maturation as his idealism is shattered by the truth of corporate reality and he awakens to fulfill his life mission and the way to true success in life. —ISBN 1-880666-04-9, $14.95

Voice in the Mirror: Will The Final Apocalypse Be Averted? by Lee Shargel. In this first novel in The Chulosian Chronicles, extraterrestrials warn planet Earth of an impending disaster, but our technology is inadequate to deal with it. The ETs must travel to Earth, but that is only the beginning of our problems.
—ISBN 1-880666-54-5 $23.95 hardcover

Transformational Tools

Intuition by Design, by Victor R. Beasley, Ph.D. A boxed set of 36 IQ (Intuition Quotient) Cards contain consciousness-changing geometrics on one side and a transformative verse on the other. The companion book tells you the many ways to use the cards in all aspects of your life to bring yourself into alignment with the Higher Mind of Source. An incredible gift to yourself or someone you love.
— ISBN 1-880666-22-7, $21.95

Navigating the '90s, by Deborah Soucek. A practical way to find safe passage through these increasingly chaotic times. Focusing on ways of freeing ourselves from our past conditioning, this book is a gentle guide toward reclaiming our true selves.
— ISBN 1-880666-28-X, $13.95

Angels of the Rays, by Mary Johanna. Twelve different angels are presented in full color to assist your healing and ascension process. Each angel has her color ray, gemstone, essence, invocation, and special message. Includes twelve removable full-color Angel Cards and directions for their use. Related tape available.
—*ISBN 1-880666-34-0, $19.95*

Bridge Into Light: Your Connection to Spiritual Guidance, by Pam and Fred Cameron. Simple, clear teaching, useful for anyone who wishes to connect with their own guidance. Offers many step-by-step exercises on how to meditate and channel, and gives ways to invoke the protection and assistance of the Masters. Companion tape available. — *ISBN 1-880666-07-3, $11.95*

In addition to the OHP titles listed above, we also offer the writings and tapes of many other inspired and inspiring authors. For a complete list of available titles and products, call for our free catalog or use the Business Reply Card at the back of this book. You'll be glad you did!

Children's Books and Tapes

Books and tapes in this category include titles such as *Mary's Lullaby,* and "The Fool Stories" book series. Although primarily intended for children and adults who interact with children, they speak to the "child" within us all.

Magical Music

We carry many titles of spiritually-based music, including both vocal and instrumental types, by artists such as Richard Shulman, Omashar, Kate Price, Lee Eisenstein, Iasos, Spencer Brewer & Paul McCandless, Ricky Byars, and Michael Hammer. A wonderful gift to yourself or a loved one! For a listing of available titles, call or write for our free catalog.

ATTENTION BUSINESSES, SCHOOLS, AND LIGHT CENTERS
OUGHTEN HOUSE books are available at quantity discounts with bulk purchases for educational, business, or sales promotional use. For details, please contact the publisher.

OUGHTEN HOUSE FOUNDATION

Oughten House Foundation, Inc. has been created as a publishing, educational, and networking organization. The purpose of the Foundation is to serve all those who seek personal, social, and spiritual empowerment. Our goal is to reach out to 560 million people worldwide. The Foundation has a non-profit (501 (c)) status and seeks members and other fund-raising affiliations. Programs for all age groups will be offered.

One of our most successful programs involves young people aged 14 - 22—the Renaissance Generation—whereby the Foundation sponsors a creativity contest and publishes the winners' entries in book form. Started in the Bay Area, we expect this program to rapidly spread nationwide.

An integral part of our mission involves the development of a global network to support the dissemination of information, especially through organized community groups. Information related to membership and program services is available upon request. Please contact Oughten House Publications, or call (510) 447-2332.

NOTE: If you have a network database or small mailing list you would like to share, please send it along!

Participate in the Divine Plan by co-creating an international network system of educational programs for Lightworkers and awakening individuals. The Foundation helps you in this process by providing and sponsoring programs through our Organized Community Groups and centers. By being involved and sharing, you can manifest planetary' change in service to the Source for the Source.

Benefits of Foundation membership include:

1. Participation in interactive correspondence

2. 10% or more discount (based on membership category) on Oughten House Publications' books and Catalog products

3. 15% discount on all lectures, seminars, and educational programs offered by the Foundation

4. Automatic affiliation with local Organized Community Groups and Centers

5. Local geographic member listings for networking

6. Special notification of upcoming events, author tours, educational programs and symposiums

7. Opportunities for membership to develop literature and publications for membership purposes and dissemination on a diversity of topics. (Topics may range from self help, spiritual enlightenment, publishing, learning the group process, communication, conflict resolution, running your own business, or any topic which serves to benefit the membership.)

8. Eligibility to join the *Networker's Program,* and become a key player in the Foundation's expansion as well as the opportunity to receive firsthand or preview materials

9. Membership is tax deductible.

Information related to membership and program services is available upon request. Please forward your inquiries to Oughten House, or call (510) 447-2372.

Publish Your Book !

Do you have a manuscript that you would like to see published in a professional way? Oughten House offers its Author Investment Program for aspiring authors who want to fund the publishing their book but are rightly wary of self-publishing and vanity presses.

Call us up to learn what it takes to get your book published. We can assist you in every aspect of the publishing process, and educate you at every step of the way.

Call **Joy Marie** at Oughten House Publications on:
(510) 447-2332 Fax (510) 447-2376,
e-mail: oughtenhouse@rest.com,
or visit our web site: www.oughtenhouse.com

CATALOG requests & Book

Catalogs will gladly be sent upon request. For catalogs to be sent outside of the USA, please send $3.00 for shipping and handling.

Book orders must be prepaid: check, money order, international coupon, VISA, MasterCard, Discover Card, and American Express accepted. Call for shipping and handling (no P.O. boxes for UPS).

To place your order, call toll-free: 1 (888) ORDER IT (888-673-3748) (orders only, please!)

For information, or to mail or fax an order, contact:
OUGHTEN HOUSE PUBLICATIONS P.O. Box 2008
Livermore California 94551-2008 USA
Phone (510) 447-2332
Fax (510) 447-2376
e-mail:oughtenhouse@rest.com
www.oughtenhouse.com

"Well Worth the growing pains!" Bill Weir Electric Cars ABC Evening News 8/11/09